Christer Bergström

LUFTWAFFE PILOTS IN WORLD WAR II:
The Veterans' Stories Volume 1

In memory of my friend Månen Lundquist

VAKTEL BOOKS

Selected previous titles by Christer Bergström
Luftwaffe Fighter Aircraft in Profile, 1997
Deutsche Jagdflugzeuge, 1999
Black Cross/Red Star: The Air War over the Eastern Front, vol. 1, 2000
Black Cross/Red Star: The Air War over the Eastern Front, vol. 2, 2001
More Luftwaffe Fighter Aircraft in Profile, 2002
Graf & Grislawski: A Pair of Aces, 2003
Jagdwaffe: Barbarossa—the Invasion of Russia, 2003
Jagdwaffe: The War in Russia January -October 1942, 2003
Jagdwaffe: The War in Russia November 1942-December 1943, 2004
Jagdwaffe: War in the East 1944-1945, 2005
Black Cross/Red Star: the Air War over the Eastern Front, vol. 3, 2006
Barbarossa: The Air Battle, 2007
Stalingrad: The Air Battle, 2007
Kursk: The Air Battle, 2008
Hans-Ekkehard Bob, 2008
Max-Hellmuth Ostermann, 2008
Bagration to Berlin, 2008
The Ardennes 1944-1945: Hitler's Winter Offensive, 2014
The Battle of Britain, 2015
Operation Barbarossa, 2016
DAISY—The History of a C-47/DC-3 in World War II and the Men who Flew it, 2019
Arnhem 1944, Vol. 1, 2019
Arnhem 1944, Vol. 1, 2019
Black Cross/Red Star: Air War over the Eastern Front, Vol. 4, 2019

The front cover image depicts the German fighter pilot Hans Philipp, II./JG 54, wearing a brown leather jacket, by his Messerschmitt 109 E. The picture was taken in France in October 1940. The rear cover image depicts a Messerschmitt 110 in Greece in 1941.

© 2018 and 2019 Christer Bergström
Cover design: Caroline Bergström
Layout: Henny Östlund
Printing: Printon Printing House, Estland
ISBN 978-91-88441-54-6

Vaktel förlag/Vaktel Books
Box 3027
630 03 Eskilstuna, Sweden
vaktelforlag.se
forlag@vaktelforlag.se

Table of Contents

Glossary and Abbreviations	4
Foreword	7
Video Clips in the Book	13
CHAPTER 1 What Was It Like to Be an Airman in the Luftwaffe? What the Veterans Recounted	15
CHAPTER 2 'Assi' Hahn—the RAF's Nemesis Meets Stalin's Son	61
CHAPTER 3 'Daddy' Mölders—the Saint of the Luftwaffe	99
CHAPTER 4 The Terrible 'Jochen' Marseille	129
CHAPTER 5 'Let's Not Talk About That'—Uncomfortable Facts	173
CHAPTER 6 Danish Pilots in the Luftwaffe	193
APPENDIX 1 German Fighter Pilots with 100 or More Aerial Victories During World War II	202
APPENDIX 2 The True Results of the Luftwaffe's Fighter Aviation	204
APPENDIX 3 The Most Successful German Fighter Wings	205
APPENDIX 4 The Most Successful Fighter Pilots Against Various Types of Aircraft	207
APPENDIX 5 Data for Some of the Most Important Fighter Aircraft During World War II	210
Sources	212
Notes	217
Index	229

Glossary and Abbreviations

Condor Legion German air force on Franco's side in the Spanish civil war.
Dornier 217 German twin-engine bomber, reconnaissance and night fighter aircraft.
"Edelweiss" The name of the German bomber wing KG 51.
Ergänzungsjagdgruppe (German) Replacement fighter group, responsible for supplying combat units with new flying crews as replacements for losses.
Fighter Group American fighter unit, consisting of 111-126 aircraft
Flying Fortress See B-17.
FuG (German) Funkgerät, radio.
Guards Fighter Regiment Soviet fighter unit that had been promoted to the honorary title of Guards Unit for outstanding accomplishments in battle.
Gruppe (German) Air group, usually consisting of three Staffel (squadron), each with 12 aircraft and one staff Schwarm (flight) with four aircraft.
"Grünherz" "The Green Heart," the name of the German fighter wing JG 54.
Hero of the Soviet Union The highest Soviet honorary title, awarded with a gold medal.
Hitlerjugend "Hitler Youth," Nazi youth organisation for boys aged 14-18.
IAD Istrebitelnaya aviatsionnaya diviziya; fighter group (Soviet), usually consisting of two to four regiments with 30 aircraft each.
IAP Istrebitelny aviatsionny polk; fighter regiment (Soviet), usually consisting of 30 aircraft.
J 88 Jagdgruppe 88, the German fighter group in the Condor Legion.
Jagdfliegerführer Fighter combat commander, German tactical commander of fighter units within a certain geographical area.
Jagdfliegerschule (German) fighter pilot school.
Fighter group (Soviet) see IAD, usually consisting of two to four regiments with 30 aircraft each.
Fighter regiment (Soviet) see IAP, usually consisting of 30 aircraft.
JG Jagdgeschwader; (German) fighter wing, usually consisting of three Gruppen with 40 aircraft each, as well as one staff Schwarm with four aircraft.
KG Kampfgeschwader; (German) bomber wing, usually consisting of three Gruppen, each with 39 aircraft and one staff Staffel with 12 aircraft.
Khalkin-Gol The battle of Khalkin-Gol was an armed conflict between The Soviet Union and Japan which took place from the spring and until the fall of 1939 by the river Khalkin Gol in the Dornod province, not far from the Mongolia-Manchukuo border. The concluding battle took place at the end of August and ended with Japan's Sixth Special Army being completely annihilated.
Knight's Cross One of Germany's highest awards for valor during World War II, considerably higher than the Iron Cross.
Knight's Cross with Oak Leaves A higher "level" of the Knight's Cross.

Knight's Cross with Oak Leaves and Swords A higher "level" of the Knight's Cross with Oak Leaves.

Knight's Cross with Oak Leaves, Swords and Diamonds A higher "level" of the Knight's Cross with Oak Leaves and Swords.

LaGG LaGG-3, Lavochkin-Gorbunov-Gudkov-3; Soviet single-engine, single-seat fighter. Was sometimes confused by the Germans with the more modern La-5 (Lavochkin-5), which was then erroneously called LaGG-5.

Leytenant Soviet military rank, equivalent of Lieutenant.

LG Lehrgeschwader; learning group, German air wing for trying out new tactics.

NJG Nachtjagdgeschwader; (German) night fighter wing, usually consisting of three Gruppen with 40 aircraft and one staff Schwarm with four aircraft each.

PVO Protivovozdushnoy oborony; home air defense (Soviet).

RAAF Royal Australian Air Force.

RAF Royal Air Force.

SAAF South African Air Force.

SB Skorostnoy bombardirovshchik (Soviet); fast bomber, name of a specific Soviet twin-engine Tupolev bomber, in Western publications sometimes erroneously named SB-2 or SB-3.

Schwarm Swarm; flight, German tactical formation consisting of four fighters.

Sergent French military rank, equivalent of sergeant.

Sergent-Chef French military rank, approximate equivalent of Battalion Sergeant-Major.

Squadron (British) air division, consisting of 12 aircraft.

Squadron Leader British air force division commander.

Staffel (German) Air squadron, usually consisting of three Schwarms with four aircraft each.

Starshina Soviet military rank, approximate equivalent of Battalion Sergeant-Major.

StG Stukageschwader (German) dive-bomber wing, usually consisting of three Gruppen with 39 aircraft and one staff Staffel with 12 aircraft each.

Stuka Sturzkampf (German); dive-bomber.

"Udet" The name of the German fighter wing JG 3.

Walrus British single-engine Supermarine biplane flying boat.

Zerstörer (German); twin-engine fighter.

ZG Zerstörergeschwader (German); destroyer wing, fighter wing for twin-engine fighters, usually consisting of three Gruppen with 40 aircraft and one staff Staffel with 12 aircraft each.

Foreword

I grew up with the Vietnam War constantly present on TV, and was affected, as perhaps most boys in that generation were, to seek out an explanation for the horrible images on the TV screen by getting interested in World War II. Since I have always also been interested in writing, the dream to write a book about this soon evolved. Since both my parents had been serving in the air force, and my home as a child was filled with aviation literature, it was quite natural that I started with the air force. The reason why it became the German air force has more to do with my personality—as a ten-eleven-year-old, I reacted against the fact that one side, the Germans, was always being portrayed as dunces in the accounts of the war in those days. I and my childhood friend Per Lindberg read Katharine Savage's youth book *The Story of World War II,* and it annoyed us that she was so blatantly partial. (This was also where my interest in questioning sources was awoken.) Per later became a non-commissioned pilot (and was forced to eject three times, I believe); spectacles by age 14 put an end to my plans of becoming a pilot.

Instead, I began reading an increasing amount of literature for adults. I would run around the local city library, borrowing stacks of books by interlibrary loans, to the despair of the librarians. Even from an early age, I used to write whole binders full of facts that I had gleaned from these books.

But what finally put me "on track" was when Peter-Paul Ophey in Merchweiler, Germany, in 1972 (where I spent a summer intending to learn German) helped me getting in touch with the German veteran airman Adolf Galland. The letter from Galland—where he invited me to his home to help me with the book on German airmen that I had told him that I was planning—was somewhat of a "no return."

Adolf Galland, the old fighter pilot general, was the uncrowned "emperor" of the veteran airmen. He kept a very firm grip of the conductor's baton. Galland opened doors for me that I had not been able to dream of. He called on his old fellow aviators everywhere to receive "der Junge" well and give him what he needed for his project. I suppose that I became some sort of an informal mascot to the old veteran airmen during those years in the 1970s.

And they did follow Galland's advice and instructions indeed. Thick photo albums were unfolded, large files of documents were opened, efforts were made to write detailed replies to the questions I posted to them. And I was just gaping and receiving. And writing.

There was intense communication between the aviators from both sides of the war, the old adversaries, and my good connections with German airmen quite automatically led to just as good connections with a growing number of Allied airmen.

I called my first book project "Die Luftwaffe." It was written in Swedish and comprised three volumes, covering the years 1939–1941, when I submitted them to the first publishers at the end of the 1970s. And I was rejected.

But one of the Grand Old Men of Swedish aviation history got wind of my work, and contacted me; he said that he could help me publish the material if he could take it with him. We met in Linköping, and he took the script and a shoebox full of photographs that I had been given by the old veterans with him. I did, however, keep all copies that I had made of documents.

Time went on, and after almost one year, I got in touch with him, and was told that unfortunately... He was going to return my material. A parcel arrived, containing the script, but no photos. "Photos? What photos?" he replied when I asked him. And then I received one of his books on aviation, "as a consolation." I suppose that is what you call learning money.

However, this elderly "gentleman" was indeed also an author of aviation books himself. He had given me a few valuable tips on how to edit the scripts to make them more readable—"you ought to insert more factboxes on individuals, more eyewitness accounts," he said, and I took that to heart.

In 1981, I moved temporarily to the town of Kalmar for studies, and there, Jan G. Andersson at the local daily became fascinated by my knowledge on World War II. He asked me to write a few articles on the matter. The first one was returned to me, filled with comments in red ink. "Janne G.," as he was called, taught me in record time how to write with journalistic drive. One year later, I was—thanks to "Janne G."—the permanent freelance writer for the media group A-Pressen of articles to the tune of "World War II—it happened today 40 years ago." In 1983, my first book was published, "Luftstrid över Kanalen" ("Air Combat over the Channel")—the result of my fervent studies of the subject, interviews with and material from veteran airmen, as well as the mentoring that I had received in the art of writing.

Since then, there have been 32 books, published in a large variety of languages, but most in English on the global market. My university studies in history led me into in-depth research in the archives, and from the 1990s on, I intensified my interview visits to the old veterans to two to three interview journeys, each at least a week long, per year. I have probably spent a total of a year of my life meeting and interviewing veterans from the war—from all sides (Germans, Britons, Americans, Russians, Frenchmen and Poles), airmen as well as ground soldiers, but most of all German aviators. I have lost count of how many veterans I have met—sometimes, for example at veterans' gatherings, there are tens of them at once—but there are definitely many hundreds of them. It has happened that I have been asked whether I have met this or that person, and then responding that I have not—only to discover later that I have several pages of interview notes from a meeting with him.

You might perhaps think that time would have outrun them by now, and most of them have passed away by now. But I am still meeting veterans.

When I changed careers in 2014 from being a high school history teacher to becoming a publisher, I finally had time to devote myself more to my interest. Then, I was able to do what I had been thinking about for a long time, to share with more people what I have gleaned from my interviews and talks with veterans from World War II, many of whom I have been able to count as my personal friends.

Indeed, I have been relatively productive throughout the years, but I will for various reasons never be able to publish all the material I have gained during my meetings with all these people. Therefore, in the spring of 214, I arranged something that I called a "Luftwaffe Weekend," where I shared things with the visitors (with words, images and film) that I had received during my meetings with German veteran aviators. This became very popular, and since then, there have been a number of Luftwaffe Weekends. This, in turn, led me to the thought that what I shared there perhaps did not have to remain within a small circle of a few dozens of participants, that I could re-edit some of my lectures at the Luftwaffe Weekends into chapters in a book. And thus, it eventually became the book on German aviators that I had told Adolf Galland in 1972 that I was planning. Perhaps I should have dedicated it to Galland, but on the other hand, my very first book was dedicated to him. Instead, I choose to dedicate this book to one of the most faithful Luftwaffe Weekend visitors, Månen Lundquist. Månen was supposed to have joined us at the latest Luftwaffe Weekend, which we held during the weekend of 27-28 January 2018, but he called from his hotel saying that he was ill and could not come. That Sunday, he was found departed for good in his hotel room. This book is for you, Månen. We will not forget you.

This book would never have materialised had it not been for the fantastic support I have received from my dear family, my wife Maria, my daughters Bambi and Caroline, my grandchildren Albin, Benjamin and Bianca, and my mother Britta.

The author and each reader of this title owes a profound gratitude to the crowdfunders below, owing to whose generous support this book series has become a reality.

Jonathan Holmes
Alessandro Voltolina
Eric Coleridge
Philippe De Meulder

Phillip Brimfield
Jeff Leach
Jesper Skejnæs
Rik Winkelaar
Lars Øyno
Timo Kilpinen
Delmar A. Davis
Michel Lemieux

I am also indebted to a large number of people who have helped me with large and small things: Alfons Altmeier, Vlad Antipov, Michael Balss, Bernd Barbas, Holger Benecke, Dag Berggren, Niklas Bergström, Dénes Bernád, Lennart Berns, Donald Caldwell, Eddie Creek, Brian Cull, Andrej Dikov, Nikita Egorov, Robert Forsyth, Carl-Fredrik Geust, Chris Goss, Ilse Grislawski, Jürgen Grislawski, Håkan Gustavsson, Nick Hector, Raimo Huovinen, Bengt Högberg, Daniel Johansson, Ivan Lavrinenko, Andrey Mikhailov, Peter Kacha, Andrey Kuznetsov, Edward McManus, Eric Mombeek, Bruce Mäki, Hans Östensson, Dr. Jochen Prien, Pernille Ravnskov, Günther Rosipal, Yuriy Rybin, Hans E. Söder, Peter Taghon, Max Thimmig, Colonel Raymond F. Toliver, Lars Tolkstam, Dariusz Tyminski, Walter Waiss, John Vasco, Manfred Wägenbaur/Traditionsgeschwader JG 52, and Peter Vollmer.

In addition, I have been using interview and/or photo materials or documents from the following veteran aviators directly for the contents of this book:

Gerhard Baeker
Günther Bahr
Cyril Bamberger
Wilhelm Batz
Petr Bazanov
Günther Behling
Helmut Berendes
Hans-Ekkehard Bob

Hugo Broch
Hansgeorg Bätcher
Hugo Dahmer
Heinrich Graf von Einsiedel
"Tim" Elkington
Hans Ellendt
Wolfgang Falck
Robert Foster

John Freeborn
Adolf Galland
Nikolaj Gapeyonok
Rudolf Gloeckner
Alfred Grislawski
Arthur Gärtner
Hans Hahn
Erich Hartmann
Hermann Heckes
Hans-Dieter Hein
Joachim Helbig
Werner Hohenberg
Klaus Häberlen
Berthold K. Jochim
Erhard Jähnert
Friitz Klees
Arkadiy Kovachevich
Sergey Kramarenko
Hans Krohn
Hans-Joachim Kroschinski
Friedrich Lademann
Friedrich Lang
Heinz Lange
Erwin Leykauf
Friedrich Lüdecke
Heinz Marquardt
Stepan Mikoyan
Victor Mölders
Hermann Neuhoff
Eduard Neumann
Friedrich Obleser
Joe Peterburs
Günther Rall
Ernst-Wilhelm Reinert
Willi Reschke
Edmund Rossmann
Vitaliy Rybalko
Heinz Rökker
Kurt Schade
Ernst Scheufele
Ernst Schröder
Gerhard Schöpfel
Johannes Steinhoff
Hannes Trautloft
Walter Wolfrum
Dieter Woratz

If I have accidentally missed anybody in the above list, I would still like to extend my gratitude.

Eskilstuna, October 11, 2019

Video Clips in the Book

Thanks to modern technology, it has been possible to illustrate this book with authentic footage from World War II. This is available through so-called QR codes.

QR codes (Quick Response) are codes for optical scanning reminiscent of common bar codes. It is easy to download software that scans QR codes on your smartphone or tablet computer. After that, it is sufficient to hold your telephone with that software running above a QR code in order for the web page with the video clip in question to be opened.

There are many free QR scanner apps that you can download onto your smartphone or tablet computer. They can be downloaded from, for example, Apple's App Store or the Android Market, depending on which type of smartphone or tablet computer you are using.

Here is the QR code for the book's internet page. There will be links there for the video clips and for some extra material.

__8. Staffel · III. Gruppe · JG 52 · Kubanbrückenkopf 1943__

Leutnant Markhoff gest.

Oberleutnant Lüty †

Hauptmann Rall

Feldwebel Schumacher †

"Rata"

Unteroffizier Hohenberg

Leutnant Puncke

Unteroffizier Lotzmann †

Auf diesem Foto fehlen:

       Leutnant Obleser — Lazarett
† Unteroffizier Hauswirth ⎫
† Unteroffizier Ahlbrand ⎬ auf Feindflug
† Unteroffizier Maiwald ⎪
† Unteroffizier Meltzer ⎭

„†" = gefallen.
„gest" = nach Kriegsende gestorben.

CHAPTER 1

What Was It Like to Be an Airman in the Luftwaffe? What the Veterans Recounted...

The image on the opposite page is taken from the German pilot veteran Werner Hohenberg's photo album, and shows a group of pilots from the German fighter squadron 8./JG 52 at the Kuban Bridgehead on the Eastern Front at the end of March 1943.

Much can be read from this image. The casual relation between officers and the rest of the men is visible without any room for doubt. Within the army, in the German navy, and in many other air forces, there was quite a strict division between officers and non-commissioned officers. In the Luftwaffe of the Third Reich—a fighting service that was no more than eight years old when this picture was taken—it was considerably more common for officers, non-commissioned officers, and even privates to associate casually between their sorties. This was especially the case among the fighter pilots on the Eastern Front, where living conditions in general were simpler. The airmen in the picture are sitting on the roof of one of the mud huts where they had been quartered at this airfield.

There were twelve pilots in the fighter squadron 8./JG 52 at this time. Four pilots—the lance corporals Wilhelm Hauswirth, Gustaf Ahlbrand, Manfred Maiwald and Karl-Heinz Meltzer—were, according to what Hohenberg has written on the picture, on a combat mission when this picture was taken. The twelfth pilot, Lieutenant Fritz Obleser, was in a hospital after being shot down and wounded on May 28, 1943. The previous day, this division had celebrated its 750th victory since the beginning of the war—a remarkable result for a unit consisting of twelve pilots. When the picture was taken, these pilots had achieved the following results for shot-down aircraft:

Günther Rall 145
Karl Schumacher 51
Wilhelm Hauswirth 49
Karl-Heinz Meltzer 28
Fritz Obleser 25
Werner Hohenberg 23
Hans Funcke 12
Hans Markhoff 4
Gerhard Lüty 1
Manfred Maiwald 1

Manfred Lotzmann and Gustaf Ahlbrand had not yet succeeded in shooting down any enemy aircraft.

This says quite a lot about the German Luftwaffe during World War II. No other air force participated in such extensive fighting for such a long time as the Luftwaffe. In this single German fighter squadron, there were thus, at this time, in the middle of World War II, three pilots—two of whom are in the picture above—who had more victories than what any of the Western Allied aces achieved during the entire war: The most successful U.S. ace during the war, Richard Bong, was credited with 40 victories, and the British top ace "Johnnie" Johnson, reached 34.[1] (The South African "Pat" Pattle, who was killed in action in Greece in April 1941, had 34 confirmed victories, but because of the retreat from Greece, many of his last victories could not be confirmed, so it is estimated that he "ought to" have had 51 aerial victories.)

The term "fighter ace" was actually only used on the Western Allied side during World War II, and then, it referred to an aviator who was confirmed to have shot down at least five enemy aircraft, which was considered an extraordinary accomplishment. In this German fighter squadron, nine out of twelve pilots were thus "aces" according to this definition. When the war was over, the final victory numbers for these pilots in 8./JG 52 looked as follows:

Günther Rall 275
Fritz Obleser 120
Karl-Heinz Meltzer 61
Karl Schumacher 56
Wilhelm Hauswirth 54
Werner Hohenberg 33
Gerhard Lüty 28
Hans Funcke 19
Manfred Lotzmann 15
Hans Markhoff 15
Manfred Maiwald 14
Gustaf Ahlbrand 5

8./JG 52 was no particular elite unit, but could be considered an average German fighter squadron at this time, at least on the Eastern Front.

The most successful fighter pilot during World War II, the German Erich Hartmann, achieved a total of 352 victories. The fighter unit with which he served, JG 52 (nominally consisting of 132 aircraft, even though the actual figure was usually considerably lower), was credited with 11,000 aerial victories—as many as the entire American fighter force in Europe and the Mediterranean. These 11,000 victories were shared between an estimated no more than one thousand fighter pilots who served at the unit at any point.

One hundred and three fighter pilots were credited with 100 or more victories each during World War II; all of them were Germans. These 103 pilots alone were responsible for a total of 15,400 victories—almost as many as all British and American fighter pilots in Europe and the Mediterranean area together during the entire Second World War: RAF Fighter Command logged more than 5,000 victories, the American fighters in Europe and the Mediterranean about 11,000. In total, the German fighter pilots reported an estimated 70,000 enemy aircraft shot down during World War II. A total of about 5,000 German fighter pilots qualified for the Western Allied designation of "fighter aces," at least five victories.

There are many reasons for the large difference in numerical results between German fighter pilots and those of other air forces during World War II. The most commonly proposed reason is that the best German pilots flew many more sorties than their opponents, and were thus much more experienced.

In the British and American air forces, airmen were almost exclusively withdrawn from frontline service after about a year or a certain number of sorties, and then let them serve as trainers. In the U.S. Air Force, such a period of service was called a *Tour*. In the Luftwaffe, there was no such system.

It has been said that in the Luftwaffe, the pilots flew until they crashed. However, this is a statement that needs qualifying; in fact, there was no other armed force during World War II where the officers showed greater care for their subordinates than in the German armed forces during World War II: the officers were taught psychology early during their training, and had the explicit task of assuring that the unit was functioning as a family for the soldiers, where the officers were to take on some sort of a fatherly role. Many times, pilots in the Luftwaffe who showed signs of fatigue were sent on "forced leave" by their superiors. There were even special resorts for aviators, where they could recuperate. It was also quite common that the most successful pilots were taken out of service because the propaganda needed them alive; but it was just as common for them to be sent back to the frontline when the state of the war had deteriorated.

Either way, there was no *Tour* system in the Luftwaffe. The aviator who carried out the largest number of missions during World War II was also a German—Hans-Ulrich Rudel, a dive-bomber pilot who flew 2,530 sorties between 1941 and 1945. That is five times as many as any Western Allied pilot achieved during the war.

What is an Aerial Victory?

An aerial victory—also called a "kill"—is used to denote when a plane has been confirmed to have been downed. The criteria for such a downing were different between the different air forces. In general, it could be said that the German and Soviet air forces had stricter requirements for a confirmed victory than the British and American air forces. In the former, a witness was required to confirm that the aircraft that had been fired at had crashed into the ground, while the custom in the latter was that an intelligence officer interviewed the pilots after combat, and assessed the results of the aerial combat from their accounts. However, in practice, the German system was not as secure as it has sometimes been claimed. It can take several minutes for an aircraft that has been shot to pieces to hit the ground, and in aerial combat with many participating aircraft and swift manoeuvres, it was difficult to keep your eyes on a crashing aircraft. A pilot is often able to regain control of a plummeting aircraft and take it back to the base, in spite of serious damages to the machine. In such cases, the crash of another aircraft, shot down by another pilot, could be counted twice by mistake.

What constituted a shooting down of an aircraft is also a matter of definition. The pilot of a badly damaged aircraft often made it by belly-landing his machine. From an altitude of thousands of metres, where the battle was taking place, the enormous dust cloud stirred up by the belly-landing could be taken for the cloud from an explosion. Here, the pilot who had been firing at this aircraft could have it credited as a confirmed victory, while the opponents could have it repaired and redeployed into combat. This is the explanation for many "unconfirmed" victories on the Eastern Front—from both sides—since aerial combat often took place above the front-line area, and both sides had their aircraft stationed at airfields only a few kilometres behind the front. Many times, a pilot managed to take his damaged aircraft up from a tailspin and carry out an ordinary landing at his own airfield.

A comparison between the combat reports from both sides show that many Allied "overclaimed" victories were due to the most common German evasive manoeuvre when the pilot was subject to an attack: an *Abschwung*—what the Americans called a *Split S:* a quick half-roll, followed by a steep dive at full throttle. During such a manoeuvre, the usually poor-quality German aircraft petrol emitted thick smoke, which could easily be interpreted as the aircraft being on fire.

Added to these circumstances was when the ground was obscured by clouds. When that was the case, there was of course no requirement that a witness had to see the aircraft that had been fired upon slam into the ground. In fact, however, both Germans and Soviets were so careful to book the true number of downed aircraft that the pilots often carried out sorties for the sole purpose of localising the wreckage of a downed aircraft.[2]

However, the German and Soviet systems also invited favouritism. Medals and fame were awaiting, and it did happen that two pilots on a mission verified incorrect "victories" for each other. Especially at two German fighter units, JG 2 and JG 5, a kind of "unit culture" of tolerance for fictitious victories seems to have appeared—this becomes clear when you compare these units' victory claims with the opponents' reported losses. This is also something that veterans from both of these units have admitted to. One possible explanation for this state of things could be that these units were stationed for long periods of time in sectors where there was not as much contact with the enemy as in other sectors, and thus not was many opportunities for achieving victories.

Even though most of the Luftwaffe's claims can be verified through their enemies' reported losses, it was not unusual that the young hotspurs simply made up victories. "We all did from time to time," the Luftwaffe veteran Alfred Grislawski admitted.[3]

There will be several examples of various reasons for "unverified" victories in the following chapters.

Indeed, Rudel was usually stationed only some ten kilometres from the frontline and could sometimes carry out as many as eight sorties in a single day. The British and the Americans usually had a longer distance to fly during their operations from the British Isles; it could perhaps be compared with the situation for the German bombers. The bomber flier (of multi-engined bomber aircraft) that carried out the largest number of sorties during World War II was the German Hansgeorg Bätcher, whose logbook records 703 sorties between 1939 and 1945. There were many other German bomber pilots who carried out more than 450 sorties:

Rudolf Müller 682
Hellmuth Kahle 526
Benno Herrmann 520
Hermann Hogeback more than 500
Karl Lipp more than 500
Joachim Helbig 480
Siegfried Röthke 476
Herbert Wittmann 467
Wilhelm Odenhardt 453
Rudolf Henne 450

On the Western Allied side, there were four bomber pilots who carried out more than 100 sorties: the British Guy Gibson held the record with more than 170 missions, followed by his fellow countryman Leonard Cheshire with 103. Both of them were awarded the highest British decoration for valour, the Victoria Cross. The American bomber pilots with the largest number of sorties in Europe or the Mediterranean both served in the same unit, the 340th Bomb Group—where, incidentally, Joseph Heller also flew missions. (Heller later wrote the novel *Catch 22*, an essentially fictitious story, which however is based on a few real-life events at the 340th Bomb Group, and where he expresses the quite harsh staff policies of the U.S. Army Air Force during the war; see *The True Story of Catch 22: The Real Men and Missions of Joseph Heller's 340th Bomb Group in World War II* by Patricia Chapman Mede.) George Wells and Fred Dyer in the American 340th Bomb Group flew 102 bombing sorties each.

These airmen reached these triple-digit figures by volunteering for several *Tours*. A *Tour* for an American bomber flier was otherwise 25 sorties—a number that was raised to 30 in early 1944, and eventually to 35.

Returning to the fighter pilots, which this book is mainly about, a few hundred sorties was the usual figure the Western Allied fighter aces reached during the war. David C. Schilling, who with 23 aerial victories was one of the most successful U.S. fighter pilots in Europe, had carried out 132 sorties when he was transferred to staff service in January 1945. While the British and the Americans replaced the veterans who had been taken out of service with newly trained pilots, the Luftwaffe kept its veterans at the front, where they developed their skills and trained newcomers.

The Germans had more experienced airmen than their opponents from the first day of the war; when World War II broke out, there was an elite of veterans who had flown in the Spanish Civil War—for example, Adolf Galland, with 280 sorties in his logbook. This advantage was one that the Luftwaffe would retain throughout the war. Even though the great losses and the shortage of fuel towards the end of the war forced the Germans to shorten the pilot training, there was a cadre of increasingly experienced veterans all the time. The French ace Pierre Clostermann spends them a few lines in his memoirs: "They knew their craft perfectly, knew all of its tricks and subtleties. They were both cautious and confident in themselves. They piloted their aircraft masterly and were very dangerous."[4]

Recklessness in combat was nothing unusual, but in general, the veterans learned to be cautious. One proverb among the fighter pilots was, "It's the enemy you don't see who will shoot you down." Therefore, they had to monitor the entire airspace the whole time. Ilse Grislawski, Alfred Grislawski's wife, recounted, "Every time Alfred was at home on leave, I used to have to mend the collar of his shirt, which was completely worn out since he kept turning his head all the time, back and forth, when he was out flying."[5] Many veteran pilots said that you could recognize the beginners on the enemy's side by the way they flew— if a formation of fighters kept the same course at the same altitude, you could be certain they were beginners. Peter Düttmann from JG

52 recounted, "You never flew straight ahead during combat missions. Willi Batz used to fly once to the right, once to the left, and kept changing altitudes. We kept looking around—upwards, back to the right, back to the left, downwards, forwards…"

On May 8, 1944, the American captain Robert S. Johnson carried out his 91st and last sortie with a P-47 Thunderbolt fighter over Germany. After that, he was sent home to the U.S., with a total of 27 aerial victories on his account. By then, he had been serving at the "front" for a total of ten months. One of the German fighter pilots taking part in the aerial battle that would become Johnson's last as the German Herbert Ihlefeld. He had already been flying in combat in the Spanish Civil War in 1937, taken part as a fighter pilot in the invasion of Poland in September 1939, and after that, he flew over France and England in 1940, on the Eastern Front in 1941-1942 and in the Home Defense of Germany from 1943. On this May 8, Ihlefeld took off twice against the American bomber formations, shooting down two B-17 Flying Fortresses, his 111th and 112th aerial victories. By then, he had more than 800 sorties in his logbook.

The American with the largest number of sorties in Europe during World War II was Donald Blakeslee, who at first flew with the RAF during 1941 and 1942 and after that, until September 1944, with the USAF. He carried out more than 500 sorties and was credited with 15½ confirmed victories. (In the American and British air forces, you shared an aerial victory when several pilots took part in the shooting down.) The most successful fighter ace of Great Britain during World War II, "Johnnie" Johnson, flew 515 missions and had 34 confirmed downings.[6] While these constituted an exception in the Western Allied side, fighter pilots with such experience were not uncommon in the Luftwaffe. Günther Rall had, when the picture at the beginning of this chapter was taken, noted 480 sorties in his logbook;[7] the number had reached more than 700 when the war was over. Herbert Ihlefeld reached more than 1,000 sorties. Hermann Graf—who was the first one to reach more than 200 victories—logged 830 sorties, Heinz Bär (221 victories) about 1,000, and Johannes Steinhoff (178 victories) flew 993 combat missions between 1939 and 1945.[8]

Three German veteran fighter pilots recount what importance their experiences had:

Günther Rall: "An unexperienced pilot is completely busy keeping his aircraft airborne, maintaining the correct number of revs, staying in touch with his flight leader."[9]

Edmund Rossmann: "My visual perception capacity improved with experience. Eventually, I could 'feel' that there were enemy aircraft in the sky—somehow, I sensed that some almost invisible change in the firmament or of a cloud formation meant that there were aircraft there."[10]

Walter Wolfrum: "We learned how to immediately see which enemy pilots in a formation that were beginners. We always started with picking them off."[11]

Alfred Grislawski, who flew 800 combat missions between 1941 and 1944, recounted that when he flew above Normandy in the summer of 1944, he was experienced enough to be able to predict exactly which maneuver his opponent in the air would do next.[12]

One important reason why the German aviators could survive so many missions at all was that their fighters—which protected the bombers—were better than most of their opponents' aircraft for most of the war, at least until the summer of 1943, which was two thirds of the length of the war. It was only then that the Russians were able to produce fighters that were equal to the German fighters Messerschmitt 109 and Focke-Wulf 190. One year previously, the British had finally caught up with the Germans' technical advantage with their new Spitfire Mark IX. The Spitfire Mark I was indeed capable of standing up to the Messerschmitt 109 E and Messerschmitt 110 already during the Battle of Britain in 1940, but by then, most of the British fighters were Hawker Hurricanes, which in total were inferior to both the German fighters. As the Germans then received their new Messerschmitt 109 Fs and Focke-Wulf 190s by the end of 1940 and 1941 respectively, they regained their technical advantage.

Of at least as great importance was the German fighter pilots' superior fighting tactics, which were developed by the aviator Werner Mölders during the Spanish Civil War. Mölders realised that the traditional three-plane V formation had become obsolete because of the fast mono fighters—it only stopped the pilots from taking full advantage of the new machines' full potential. The V formation had been invented during World War I to facilitate visual contact between the pilots, but now, there was radio contact between the aircraft.

In combat, the V formations tended to scatter, simply because even a simple turn at the same speed made the aircraft disperse immediately, so that it was every man for himself. But Mölders advocated teamwork as no-one before within fighter aviation: the basic tactical force would be a pair of aircraft staying together.

The pilot on one of the aircraft was the flight leader, and was to operate offensively, to take down the enemy aircraft. The pilot of the other aircraft, the wingman, was tasked with protecting his leader. During the approach, ahead of combat, two two-ship formations (*Rotte*) were joined together into a group (*Schwarm*), helping each other to scan the skies. Since the two two-ship formations in the group flew quite far apart, they became more difficult to spot at a distance than a closely connected three-plane formation. The veterans used to say, "Discovering the enemy before he discovers you is half the victory." When there was eventually contact with the enemy, the group was dissolved into the original two-ships.

When World War II began, the British and the French flew in the old V formations, and the Russians did the same in 1941, when Hitler began the German-Russian war.

The very young Luftwaffe—only formed in 1935—recognized fresh thinking and questioning old dogmas. Its opponents, however, were inhibited by conservative thinking. The American air force played a completely decisive role during the invasion of Normandy in 1944 as well as during the following advancement by its close air support for the ground troops, but the fact is that it took many years for the conservative American air staff to take in the lessons learned by the British concerning the importance of coordination between the air force and ground troops. The tactic for this had been developed by the British in North Africa in 1941-1942, and they had in turn learned it from the Luftwaffe—whose close air support with Stuka dive-bombers was a cornerstone in the so-called *Blitzkrieg* ever since 1939.

The Soviet air force was especially conservative at the beginning of the war with Germany. New thinkers such as the fighter ace Aleksandr Pokryshkin were tearing their hair over the tactical dogma which dictated "horizontal air combat" (that is, no climbing or diving). The Soviet air force officers argued that the pilots should avoid losing altitude in aerial combat, since they argued that that put the at a disadvantage against an enemy being at a more elevated position. This was purely academic thinking.

The German fighter pilots, on the other hand, mainly applied what the Russians called "vertical air combat": when they discovered the enemy, they sought—as far as was possible—to place themselves higher than him with the sun in their backs, or hidden behind a cloud. After that, they carried out a dive attack, fired at the enemy at as close a range as possible, and then climbed to a higher altitude again. By then, the enemy formation had often become dispersed through the shocking attack, and the German fighter pilots could, thanks to the speed surplus that they had accumulated during the dive, climb towards the sun only to repeat the whole procedure soon afterwards. It would take some time before the Luftwaffe's opponents began making a system of this tactic.

Another important factor behind the success of the German fighter command—and indeed also the bomber and fighter-bomber commands—was the offensive way in which the fighter pilots were being used. Hermann Göring, the founder and commander of the Luftwaffe, was himself a veteran fighter ace from World War I. He had been flying together with Manfred von Richthofen, the "Red Baron," who, with his 80 victories was the most successful fighter pilot of World War I. Von Richthofen was completely focused on the offensive—he was only there to down the enemy's aircraft, nothing else, and he did this as if it had been some kind of sport; he even called his sorties "a noble form of human hunting." Von Richthofen and many of the fighter pilots of World War I saw aerial combat as a sort of a renaissance for the tournaments of the Middle Ages, where brave men met each other in duels high up in the air above the muddy trenches. "Find your enemy and shoot him down—all else is unimportant," was the maxim of the "Red Baron."

Hermann Göring, who had grown up in an old medieval castle that had spurred his imagination, took this to heart wholeheartedly. He impressed this attitude onto his young fighter pilots, who—even though some of the want to claim the opposite—actually were his favourites.

The fights and duels in the air during World War I had been decided by the relatively short ranges of the aircraft—by the time they reached the airspace above the front, there was little else the fighter pilots could do than seeking and destroying the enemy's aircraft. However, with the development of more modern, multi-engine bombers with longer range and greater payloads, the fighter command began having an increasingly defensive task. At least during the first few years of the war, the fighter commands of Germany's opponents were defensively orientated, with formations of fighter planes passively circling the air, waiting for German aircraft to turn up.

Göring's fighter pilots, on the other hand, were being used offensively. So-called free hunting—"find your enemy and shoot him down"—would remain their main task for most of the war. One important aspect of this, which is often overlooked, is that this also gave the fighter pilots an opportunity to choose whether or not to engage in combat. During a free-fighting mission, they could avoid a larger, and thus more dangerous, formation of enemy fighters, and instead attack a smaller group. On the other hand, those who waited for the enemy to turn up, had to take the fight regardless of the odds.

The purpose of this description is—far from idealising the air force of the Third Reich during World War II—to provide one of several perspectives of the Luftwaffe, the history of which is completely unique in aviation history. With the exception of the submarine crews, mortality was not higher anywhere in the German Wehrmacht than among the aviators, especially the fighter pilots.

When the image at the beginning of this chapter of the pilots of 8./ JG 52 was taken, Gerhard Lüty had recently (on May 21, 1943) barely survived being shot down himself. During the last three days of March 1943 alone, four of the pilots of 8./JG 52 were shot down, but escaped unharmed. One year after the picture was taken, half of the pilots that the division had had by the end of May 1943 had been killed. In one single day, on July 5, 1943, both Lotzmann and Hauswirth were shot down and killed, while Schumacher was shot down and wounded. Lüty was written off as missing after a mission on July 8, 1943, Meltzer was killed in 14 August 1943, and Ahlbrand was written off as missing on August 22, 1943. On January 7, 1944, Maiwald himself was shot down and killed. Schumacher was killed in aerial combat on May 31, 1944—the third time he was shot down in ten months. Moreover, by then, since the picture was taken, twelve of the young pilots that 8./JG 52 had received for their losses had been killed:

5 July 1943 Lance Corporal Martin Lesckowitz
7 August 1943 Corporal Karl Reich
11 August 1943 Lance Corporal Heinrich Mühlschwein
21 August 1943 Lance Corporal Kurt Rathmann
22 August 1943 Lance Corporal Günter Münschow
10 October 1943 Lance Corporal Werner Pohrt
12 October 1943 Lance Corporal Otto Mixa
13 October 1943 Lance Corporal Siegfried Tenner
23 March 1944 Lance Corporal Karl Kinnen
15 April 1944 Lance Corporal Karl Munkert
18 April 1944 Lance Corporal Willi Laggai
22 May 1944 Lance Corporal Heinz Ullmann

Out of the airmen of 8./JG 52 in May 1943, only Rall, Obleser, Hohenberg, Funcke and Markhoff survived the war. Out of them, I have met and interviewed three—Rall, Obleser and Hohenberg. When I asked them how it felt to see how little chance they had to survive during the war, I had the same answer: We were young and thought we were immortal.

At the outbreak of World War II, there were about 800 German fighter pilots. The chances of surviving the war is visible from the fact that 8,500 fighter pilots were killed and 2,700 were written off as missing between 1939 and 1945. Of course, the chances of survival varied during the war.

"Turnover time"—that is, how long it took on average before a unit had had losses inflicted corresponding to 100 % of its total strength at any one moment—was, for the fighter command on the Western Front:
- 1940: More than 2 years
- 1941: 2 years
- 1942: Less than 2 years
- 1943: 10 months
- 1944: 6 months
- 1945: 6 months

"Turnover time" for the fighter command on the Eastern Front looked slightly better:
- 1941: 2 years
- 1942: 2½ years
- 1943: 1 year 4 months
- 1944: 1 year 3 months
- 1945: 6 months

A common erroneous conclusion of this is that the Soviet aviators were worse than the Western Allied. That was generally not the case. It is true that the Soviet air force, due to its very heavy losses in 1941 and 1942, was forced to drastically cut the pilots' training time to cover the gaps in the frontline units. Therefore, a large number of insufficiently trained pilots were deployed at units, where they were often easy prey for the German fighter pilots. But this was also true for the British air force in 1940-1941. As we shall see in the chapters about "Assi" Hahn and Hans-Joachim Marseille, the success of the German fighter command in combat with the RAF in 1941-1942—calculated as the quota between downings and own losses—was fully on par with the successes on the Eastern Front.

"Turnover time" for the German fighter command was the same on the Western Front as on the Eastern Front during 1941. The fact that it would later become somewhat longer in the West during 1942 is mainly down to the fact that fighting intensity by then was higher on the Eastern Front. When Paul Galland (Adolf Galland's younger brother) was killed in combat over the English Channel on October 31, 1942, he had been flying less than 100 combat missions since June 1941. During the same period, Johannes Steinhoff carried out 600 combat missions on the Eastern Front.[13]

The prospects of survival fell both on the Eastern and Western Fronts during 1943, but the downturn was more dramatic on the Western Front. This is mainly explained by the Luftwaffe being drive on the defensive on the Western Front, where fighter pilots instead of free fighting were deployed against large and dense formations of American four-engine Flying Fortresses, where each aircraft was armed with a dozen heavy 12.7mm machine guns.

German aviation historians Hans Ring and Werner Girbig provide a telling example of what it might have looked like when the small German fighters clashed with these Flying Fortresses: on May 21, 1943, fighter group I./JG 27 took off with nine Me 109s from the Dutch airfield at Leeuwarden, and attacked a formation of B-17 Flying Fortresses on their way in to bomb Emden. During the first frontal attack, Lieutenant Josef Jansen shot down a B-17, but the engine of his own aircraft was hit by the bombers' defensive fire, so Jansen had to abort and carry out a quick emergency landing, upon which he injured his head. During the following fifteen minutes, the fighter pilots Georg Stabnau, Walter Saynisch, Horst Grimm and Helmut Beckmann closed in on the Flying Fortresses to the point where they were able to shoot down one B-17 each—or at least claim to do so. (On that day, the German fighter command reported the downing of 22 four-engine bombers, while the American reports show that 14 were lost.) However, out of these, only Helmut Beckmann returned to base for a normal landing. Stabnau was forced to belly-land on the beach at Borkum, Saynisch had the fuselage of his Me 109 shot to pieces, and Grimm was forced to make an emergency landing in the Dutch countryside. Another Messerschmitt 109 that had had its entire tailfin shot off made an emergency landing and was so badly damaged that it had to be sent back to the factory in Germany for extensive repair work, yet another one made an emergency landing in Germany and was just as badly damaged, and one 109 made an emergency landing in Leeuwarden. Out of nine Me 109s that had taken off in Leeuwarden by noon on May 21, 1943, only two returned to base reasonably undamaged.[14]

From 1944, the German fighter pilots in the West were transformed from hunters into prey; the American air force had finally learned the free-fighting tactic, and with superiority in numbers many times over and technically superior fighters—especially the P-51 Mustang—they actively sought out the German fighters that were being deployed against what were now even larger formations of Flying Fortresses. It is not hard to imagine the result when great numbers of aggressively attacking American fighters were added to the difficulties that have been described above during the German fighter pilots' attacks on four-engine bombers.

With a "turnover" of six months during 1944—in other words, the entire German fighter command in the West was basically annihilated two times over during the course of twelve months—the quality of the fighter units in the West was worn down very quickly. On the Eastern Front, losses did indeed rise as well, but there, the fighter pilots had much greater chances of survival during 1944. Günther Rall, who was transferred in 1944 to the Reich Air Raid Protection League in Germany, where he was shot down and wounded, said, "The main reason why we had greater losses in the West were the relative strengths." To this, two other explanations can be added:

1. On the Eastern Front, the main task of the Soviet fighter pilots was defensive, to protect their own bombers and fighter-bombers, and missions for free fighting were exceptions. This had a natural explanation: due to the dwindling number of German aircraft on the Eastern

Front, and a shortage of fuel on the German side, the Luftwaffe was becoming an increasingly rare sight on the Eastern Front, which was why fighting of German units on the ground became the most important task of the Soviet air force.

2. While a large number of Luftwaffe veterans were killed in the great aerial battles against American bombers and fighters—in formations of several hundred at a time—the veterans on the Eastern Front accumulated even more experience during free fighting.

In fact, the average German airman on the Eastern Front in 1944 was better than ever before on the Eastern Front —or on any other sector of the front. Every tenth pilot within JG 52 on the Eastern Front in May 1944 was an ace with more than 100 victories!

The notion that the Soviet air force would somehow have been of a lesser quality than the Western Allied air forces is completely unfounded. The most successful fighter aces on the Allied side during World War II were Soviet: Ivan Kozhedub with 62 and Aleksandr Pokryshkin with 59 victories.

What, then, was the situation and life like for a German airman during World War II? What do we learn from what the veterans have to tell us? Describing that is the main purpose of this book. In spite of hundreds, perhaps thousands of books having been written on the German Luftwaffe during World War II, hardly anybody has attempted to seriously get to grips with the airmen. How did they live? What were they thinking? Which beliefs did they have? What did their everyday lives look like (because there was actually everyday life at the front as well)? In order to get some answers to this, I will let the veterans speak for themselves.

To begin with, we need to look at how you would become a Luftwaffe airman. It was mainly a voluntary application, and a strict selection was made. After completing basic military training, the hopefully prospective pilot was sent to a school of aviation. The flight training was divided into, first, a so-called A School, and, after that, a B School, where the difference was the types of aircraft that were flown.

In August 1939, the 20-year-old Luftwaffe recruit Alfred Grislawski was accepted at the Delmenhorst flight school in Germay. There, he received 480 training flights, mostly take-offs and landings, totalling about 80 flight hours.[15] After receiving his pilot's licence, the fresh pilot was transferred to a special school of aviation for the kind of military aircraft that the pilot was considered suitable for. This selection was based on a psychological test of suitability. This could be hard enough. As we shall see later, the fresh pilot Hans-Joachim Marseille received the following assessment:

Bomber pilot—not suitable

Transport pilot—not suitable

Reconnaissance pilot—not suitable

Fighter pilot—"excellent, providing that he gets an understanding commander."[16]

Alfred Grislawski was transferred to a fighter pilot school. There, he flew an Me 109 for the first time on April 25, 1940. After 34 flight hours of training in fighters—half of which in Me 109s and the other half in the Czech single-engine biplane Avia B 534 and the twin-engine Focke-Wulf 58—he graduated as a fighter pilot and was sent to a so-called replacement group, Ergänzungsjagdgruppe Merseburg. An Ergänzungsjagdgruppe was a unit where newly trained fighter pilots had their training refined by more experienced combat pilots. Here, Grislawski received another 456 minutes of flight training in Me 109s before finally getting his first deployment in August at a fighting unit, III./JG 52.[17]

Bomber pilots, reconnaissance and transport pilots also attended a C School with more advanced navigation and instrument flying before being sent first to, for instance, a bomber pilot school and then to a replacement bomber group, Ergänzungskampfgruppe, in the bomber pilots' case. German bomber ace Hansgeorg Bätcher was clear about his opinion, "Those who were no good for heavy aircraft had to become fighter or Stuka pilots." The fact was that the German bomber pilots maintained a very high standard of training throughout the war. Comradeship was also usually greatest within a bomber crew, where the personnel used to work in close collaboration in a completely different way than within the fighter command.

However, comradeship within German aerial units during the war were just as important as within all primary groups during a war. Contributing to this were the officers' training, which placed special emphasis on creating primary groups with strong fellowship, and actually also the propaganda. The ideologically impregnated states of Germany and the Soviet Union understood, in a completely different way than Great Britain and the USA, motivating their soldiers with the notion that their struggle had a "higher purpose." This had an impact that has been underrated in historiography, not least because of German veterans' efforts—at least in public—to strongly renounce the Third Reich in its entirety after the war. But the usual error to study history through the eyes of the present day also makes it easy to dismiss the effects of propaganda.

Now, the average German airman may not have lived up to the Hollywood cliché of the "fanatical Nazi," but it was not in that respect that the propaganda had an impact on these young men either. Throughout all history, only rather few people have been thoroughly ideological, and that was also the case in Nazi Germany during World War II. But the propaganda, deeply insightful of psychology as it was, influenced the young soldiers' attitudes, and it shaped their self-image. By creating an image in books, newspapers and movies of what the Soldier/Airman was like, a model was hammered out that the German teenagers took to heart. Then, when they had received their "wings" and were deployed, they often acted according to the model that they had learned. This is a social phenomenon that we can observe among today's "brats," or—at the other end of the social scale—among youngsters from deprived areas being influenced by society's expectations, manifested in movies and other mass media.

In order to penetrate the average German airman's mind during World War II, one has to lay aside all of today's values and attitudes and look at the values, norms and attitudes in Central Europe in the 1930s and 1940s. The spirits within the entire Luftwaffe were generally very high throughout the war, from the beginning to the end.

10:54-11:15 in the video clip shown below shows a group of Luftwaffe pilots. Veterans whom I have spoken to confirm that it actually reflects the true mood among German air units.

The effect of Goebbels's propaganda is proven by the fact that it still influences people, 75 years after the war. Even today, interest in World War II remains great, and those of us working within the publishing business know that it is especially the German Wehrmacht during World War II that is fascinating people. What is this, if not a remaining effect of Goebbels's very cleverly conducted propaganda?

Another effect of this propaganda was the sense of superiority over all opponents that it ingrained in the German soldiers, especially the airmen. "We felt so superior to all of our enemies. We thought that 'nothing can happen to me,'" said Hans-Ekkehard Bob, German fighter pilot 1939-1945.[18]

Specially composed, rousing marching music with lyrics appealing to the young airmen's vanity was ceaselessly being played. In his book *Psychological health effects of musical experiences: theories, studies and reflections in music health science* (2014), the stress researcher Töres Theorell at the Karolinska Institute presents research showing how e.g. modern "gym music" can make us perform better physically than without music. Marching music has a well-known stimulating effect on aggressive emotions and willingness to act, especially when combined with lyrics leading in the same direction. In his memoirs, the German fighter pilot Adolf Galland especially mentions the march given the name *Bomben auf Engelland* [sic] ("Bombs over England"):
"From loudspeakers all over the Greater German Reich, from Aachen to Tilsit, from Flensburg to Innsbruck, yes, over the army radio in the occupied countries all over Europe, the song rang out, 'Bomben auf En-ge-land'. By heavy use of kettledrums and a murdering rhythm,

accompanied by engine noise, they tried to accomplish a mass psychological effect."[19]

In his memoirs, characterised by post-war afterthought and self-righteousness, Galland throws in a reservation that "this song in no way corresponded to the severity or the reality of our fighting, just as little as did the repulsive chorus of the so-called fighter pilot song: '... *die stolze Maschine, sie wackelt, wackelt*' ["the proud aircraft, it's rocking, it's rocking—the manoeuvre that Germans did with the wings of their aircraft to indicate a downing upon return to base]." In order to understand the reality that Galland is speaking about here, it is important to keep a critical attitude to the public statements—especially in the shape of memoirs—that German veterans have made after the war.

In reality, hardly anybody could embody the "Fighter Pilot Song" (originally *Es blitzen die stählernen Schwingen*; "There's Lightning From the Wings of Steel") more than Adolf Galland, who had completely taken to heart Göring's and von Richthofen's attitudes during the war about collecting air victories as a kind of hunting trophies.[20] During this particular time, the Battle of Britain, he was a strong competitor with Werner Mölders (see the chapter on Mölders) for the position as the most successful fighter ace in the Luftwaffe. On one occasion, he told Mölders that he wanted to become "this war's Richthofen"—upon which Mölders replied, "Then I want to be this war's Boelcke!" (Oswald Boelcke was another of the truly great German fighter aces of World War I.) What is the choice of von Richthofen and Boelcke influenced by, if not one of the verses in the "Fighter Pilot Song":

So jagen wir kühn und verwegen,
In treuer Kam'radschaft verschweißt.
Der Sonne, dem Siege entgegen,
In Boelckes und Richthofens Geist.
– So we're hunting, boldly and bravely
welded together in true comradeship.
Towards the sun, towards victory,
in Boelcke's and Richthofen's spirit.

"The Fighter Pilot Song"

Bomben auf Engelland was composed by Norbert Schultze (who also composed the perhaps most famous hit of World War II, *Lili Marleen*), originally as a musical theme for the Nazi propaganda movie *Feuertaufe—der Film vom Einsatz unserer Luftwaffe im polnischen Feldzug* ("Baptism of Fire—the Film About our Air Force's Efforts during the Polish Campaign") from the winter of 1939/1940. Then, it was named *Bomben auf Polenland* (Bombs over the Land of Poland), but, ahead of the Battle of Britain, it was renamed *Bomben auf Engelland* and was given new lyrics.

By illustrating *Bomben auf Engelland with clips from scenes with German airmen from the movie Battle of Britain (1969),* the YouTube signature "LeMarTV1" has enhanced the atmosphere within the Luftwaffe during the summer of 1940 even further. Watch the video through this QR code:

Adolf Galland himself served as an advisor for the movie, so the portrayal of the German airmen is presumably quite authentic. The atmosphere and the dialogue during the orders to the German fighter pilots between 00:28 and 00:42 in this scene from the same movie would hardly have existed had Galland not had a finger in it:

The unit commander is briefing the pilots: "The mission is to annihilate the RAF on the ground."

"But then, there's nothing left for us," one of the fighter pilots objects light-heartedly. But the unit commander quickly collects his wits, saying, "Don't worry, gentlemen, the bombers won't be able to destroy everything on the ground. There'll be a few Spitfires left for you—even for you, Bruno!"

This scene, as does the photograph at the beginning of this chapter, illustrates the relatively relaxed atmosphere between officers and privates in the Luftwaffe, especially within the fighter command. Walter Wolfrum, one of the fighter pilots on the Eastern Front, said, "We fighter pilots were true front swine. The bombers, who were stationed further to the rear, were more posh gentlemen!"[21]

Indeed, there were examples of the opposite too—in the following chapters, we shall see a few examples of this—but this being considerably more common within the Luftwaffe that within any other air force is a strong impression after encounters and interviews with a large number of veteran pilots from Germany, Russia, the United States, and Great Britain during five decades.

Heinz "Esau" Ewald recounted what happened when he, as a newly graduated fighter pilot in 1943, reported for his first frontline duty to the commander of the II./JG 52 wing on the Eastern Front, Captain Gerhard Barkhorn. By then, Barkhorn was an ace with more than 100 victories and carried the Knight's Cross with Oak Leaves. When Ewald had reported, Barkhorn asked four questions in quick succession:

"*Können Sie fliegen?*" (Are you able to fly?)—Jawohl, Herr Hauptmann!

"*Können Sie schiessen?*" (Are you able to shoot?)—Jawohl, Herr Hauptmann!

"*Können Sie dranbleiben im Luftkampf?*" (Are you able to maintain your position as wingman during aerial combat?)—Jawohl, Herr Hauptmann!"

"*Können Sie sterben?*" (Are you able to die?)

– Jawohl, Herr Hauptmann…[22]

By this time, Captain Barkhorn was twenty-four years old and "Esau" was twenty. Thus, it was very young and often quite frivolous boys who were conducting this war and who developed into what the Russians called the German aces, "Terrible flying wolves." Ewald survived the war with 84 aerial victories, Barkhorn became, with 301, the second most successful fighter pilot of World War II (after Hartmann with 352).

Another episode that "Esau" recounted from his life as a fighter pilot on the Eastern Front reflects both this frivolousness and the feeling of superiority that the German fighter pilots had. On one occasion, "Esau" was flying as a wingman for another one of the great aces within JG 52, Willi Batz. At the airbase, the ground crew used to overhear the fighter pilots' radio calls, and now, the following dialogue was being heard between the two airmen—during an aerial battle with Soviet Yak fighters:

Esau: "*Vorsicht, Vorsicht, Radetzky 1, hinter Ihnen sitzt einer!*" (Careful, careful, Radetsky 1, one of [the Russians] is right behind you!)

Batz: "*Nur Mut, nur Mut, beruhige dich nur. Hinter dir sitzen die anderen drei!*" (Take heart, take heart, just calm down. Behind you are the other three!)

This Willi Batz won 237 aerial victories during 445 sorties during World War II, and survived. He accomplished one of the largest daily results for a fighter pilot during the war by downing 15 Soviet aircraft on May 31, 1944 during seven missions over Iasi. When I met Batz at the end of the 1970s, his calm self-confidence as a fighter pilot was still visible. When I asked, being young and eager (I was not even 20 years old by then) how he had been able to shoot down 15 aircraft in one single day, he replied, "Oh, I was just lucky. I was flying at full throttle all the time and one Russian aircraft after another just turned up in front of me."[23]

Now, the so-called "Air Battle for Iasi" in Romania on May 30-31, 1944 was no easy task. It cost Barkhorn's and "Esau's" fighter group II./JG 52, one third of its supply of aircraft. Since the aerial battles took place close to the German airfield, many pilots made it by a quick emergency landing, but among those who were shot down was Gerhard Barkhorn. His injuries were so severe that he did not return to the unit for half a year. It was also during this aerial battle that the 8th squadron's Karl Schumacher was killed. Had it not been for the veterans, it would have been much worse: out of the 25 claims for shot down Soviet aircraft that III./JG 52 submitted, the three veterans Batz, Erich Hartmann and Birkner were responsible for 21.

But the mood among the fighter pilots remained good, and it was not least down to the attention that they earned in the propaganda of the Third Reich. No group of soldiers during World War II have reached such great success and been so highly decorated as the fighter pilots of the Luftwaffe.

When World War II broke out, Hitler had had a new highest decoration instituted, the Knight's Cross. Germany's previous highest decoration, Pour le Mérite, was, as was the practice in many other countries, only intended for officers, but the Knight's Cross was awarded regardless of military rank. Out of 12.5 million armed men in Germany during World War II, the fighter pilots only constituted 0.2 per cent, but out of 7,000 Knight's Crosses that were awarded, almost 600 went to them. Eventually, new and higher degrees of the Knight's Cross were instituted: the Knight's Cross with Oak Leaves in 1940, where four of the first five recipients were fighter pilots; the Knight's Cross with Oak Leaves and Swords in 1941, where the first four recipients were fighter pilots; the Knight's Cross with Oak Leaves, Swords and Diamonds in 1941, where the first five recipients were fighter pilots. It was not until March 1943 that the "Desert Fox" Erwin Rommel became the first non-fighter pilot to be awarded the Knight's Cross with Oak Leaves, Swords and Diamonds.

We have seen what achievements that many German bombers made—several of them carried out four, five hundred, and upwards of seven hundred sorties over enemy territory. However, among the German bombers, none was awarded the highest decoration for valour of the Third Reich, the Diamonds for the Knight's Cross with Oak Leaves and Swords. Three bombers and six dive-bombers were awarded the Knight's Cross with Oak Leaves and Swords; among the fighter pilots, thirty-four received this high award, and nine of the were awarded the Diamonds. One important reason for this is of course the fact that the commander of the Luftwaffe, *Reichsmarschall* Hermann Göring, was a veteran fighter pilot himself.

The favouring of the fighter pilots did of course arouse a certain degree of jealousy among the other airmen. During the war, there was a satirical "description of the future" of an imagined victory march after the expected final victory of the war circulating among the Luftwaffe's Stuka pilots. The headline reads "The Entry Through the Brandenburger Tor in 1961," and it describes, more or less picturesquely, the following scene:

- First comes the World Marshall, dressed in jingling gold, and to the music from 175 Luftwaffe orchestras.
- Then follows the Half World Marshall, followed by the Sub World Marshall.
- Cheered by the crowds, the people's darlings then follow, the fighter pilots—tanned, used to victory, with the Giant Cross of the Knight's Cross on a motorised cart, complete with musical boxes and fireworks.
- After long, a stooping, greyed woman then follows, with a sign: "I am the widow of the last reconnaissance pilot."
- Next follows a man with white hair and a top-secret binder, on which it says, "In memory of the bomber fliers."
- Then nothing happens for a while, and people start to go home.
- Then, a chariot rolls forth, drawn by four horses. On the chariot there is a cage with a man in chains. It is the last Stuka pilot. Above the cage, there is a sign: "Warning! Do not provoke! Could be dangerous!"[24]

But morale was generally very high among the airmen in the Luftwaffe. "We always believed in victory," said the Stuka pilot Erhard Jähnert.[25] The fighter pilot Alfred Grislawski recounted that "when the newcomers towards the end of the war were asked for volunteers for a dangerous mission, they all raised their hands."[26]

This was probably as much down to the propaganda—which affected the general attitudes more than the opinions per se—as to the youth of the airmen. Helmut Berendes, a bomber who flew, amongst other aircraft, the four-engine Heinkel 177s, said, "I was young and wanted to fight. To us, it was all about the flying, not Hitler."[27]

To this was added the masculinity and virility that the Third Reich created for the Luftwaffe airmen. Their elegant blue uniforms—specially designed by Hugo Boss—already made them attractive to the opposite sex. When the propaganda then made the Luftwaffe airmen into sheer "superstars," they had scores of what today is known as "groupies." After all, sex is one of mankind's most important drives, and when success in combat increased the young aviators' sex appeal, that spurred them on even further.

Nazism is supposedly extremely puritan and conservative, but in essence, it is not a proper ideology, but rather a hotchpotch of opportunist ideas that are often mutually contradictory. Especially the Luftwaffe airmen were able to live a life of much debauchery. At home in Germany, they were able to have as many young ladies as they liked, sometimes with the help of the personnel department, such as at the "Fighter Pilots' Home" at Wiessee, which was a kind of a rest home for fighter pilots. At the front, things were arranged in such a way that there was an ample supply of prostitutes. Among the veterans, "jokes" are still being told reflecting frequent association with prostitutes.

Several of the airmen's nicknames stem from their sexual habits, as for example "Graf Punski," which could be translated directly as "Earl Fuck." Many of the painted personal emblems on the aircraft are also about sex, sometimes quite crudely. Many model-makers have built a Focke-Wulf 190 D-9 and put on the sticker reading *Rein muss er, und wenn wir beide weinen* perhaps without considering its meaning. The German translation is, "It's going in, even if both of us are crying."

One of the veterans I have met has also spoken about erotic relations between men in the Luftwaffe. This was something that happened—in spite of it being something that could send you to a concentration camp during the Nazi era—and, according to this veteran, it was something that everybody kept quiet about in the name of comradeship.

Indeed, there were those who were less motivated for combat, even within the Luftwaffe. According to the old fighter pilot Hugo Broch, "80 % were motivated and 20 % were not motivated, but you didn't betray anyone."[28] The veterans recount how some of their comrades, and sometimes themselves, for various reasons sometimes stayed away from combat missions. This often followed some traumatic experience, such as being shot down, or it was because they were so exhausted that they could not take any more. In total, however, the Luftwaffe units were characterised by high morale and a sense of supremacy towards the enemy right until the end of the war.

An excerpt from the book *The Ardennes 1944-1945: Hitler's Winter Offensive* by Christer Bergström:

"During the German air raid on the American Metz-Frescaty airbase on 1 January 1945, the Messerschmitt 109 'White 11' from IV. Gruppe/Jagdgeschwader 53 with Oberfeldwebel Stefan Kohl at the controls was shot down by ground fire and crashed near the airfield. Kohl tried to get back to the German lines on foot, but was arrested by French resistance fighters and taken to the air base. There he was interrogated by Major Robert Brooking in the 386th Fighter Squadron's command post barrack. Brooking had led the crucial air strike against SS-Kampfgruppe Peiper at Cheneux on 18 December, for which he received a Silver Star. But now little more than smoldering scrap heaps remained of many of the aircraft that had performed this famous mission.

Even though it was the German pilot who was in captivity, he appeared in a merry and superior manner, almost as if the roles would have been reversed. While Brooking asked his questions, Oberfeldwebel Kohl suddenly got up from his chair, walked to the window, pointed with one thumb at the rows of bullet-riddled and still smoldering Thunderbolt planes and said with a broad smile in perfect English, 'What do you think of That?'"

The Luftwaffe airmen were really living in a "bubble." This is also apparent from the recordings that exist of the secret monitoring of the German prisoners of war that the British intelligence made during the war. The German historian Sönke Neitzel and the social psychologist Harald Welzer have gone through this material and compiled some of it in their famous book *Soldiers: German POWs on Fighting, Killing, and Dying*. They write, "Surprisingly, though, the protocols do not bear out the idea, postulated by historians, that German fighting morale declined toward the end of the war. Airmen who were shot down in 1945 do not talk more frequently about being afraid to die than those captured earlier. Instead, they still proudly recount their triumphs."

On March 11, 1945—less than two months before the end of the war—Second Lieutenant Antonius Wöffen from the JG 27 fighter wing was shot down and captured. In a prison in England, he met Lieutenant Hans Hartigs from fighter group JG 26. He had been shot down and captured on December 26, 1944. The tapes from the eavesdropping by the British intelligence of the cell where the two had been placed contains the following dialogue:

> Hartigs: "Is the combat spirit among the pilots still high?"
> Wöffen: "Of course. We hope that the fortunes of war will turn soon!"[29]

Many young pilots really enjoyed the aerial combat. Hans Philipp, a fighter pilot who fought both during the Battle of Britain and on the Eastern Front told a comrade, who wrote it down in his diary, the following during the war, "A skirmish with a Spitfire over the Channel or a Yak over Leningrad is really fun!"[30]

Erich Hartmann, who reached 352 victories, more than anybody else, recounted, "Aerial combat was really exciting. You were seized with hunting fever. The sound of the engine at full throttle, the enemy aircraft that came hurtling towards you as you dove, the smell of gunpowder—all of that made the adrenaline pump. And then the reception at the airport when you came home with a new downing. It cannot be described."[31]

Of course, the brutalising effect that war has on people plays an important role here. The effect of this on the young airman is clear from several interviews in the book *Soldiers: German POWs on Fighting, Killing, and Dying*.

Lieutenant Pohl, a Luftwaffe bomber, was overheard saying in British captivity, "On the second day of the war, we were ordered to drop bombs on the railway station in Poznan. Eight of our sixteen bombs fell among the houses in the town. That didn't feel very good. On the third day of the war, something similar was repeated and then I didn't care one whit. On the fourth day of the war, I thought it was fun. I felt sorry for the horses that we were killing with our bombs, but not the people. The horses' dying screams drowned out the sound of the engines of our aircraft."[32]

Another German airman is quoted in the same book: *"When we had dropped our bombs, we fired away with our machine guns. We were shooting at everything we saw—cows, horses, it didn't matter. We were firing on trams and everything. It was great fun."*[33]

Lance Corporal Oswald Fischer in the fighter-bomber squadron 10.(Jabo)/JG 26 managed to fly 31 fighter-bomber missions—that is, dropping bombs with a fighter —over England before being shot down and captured on May 20, 1942. In captivity, he was overheard saying this to a fellow inmate who was also a German airman, "I killed lots of people in England! We were under orders to drop our bombs among the houses in Folkestone. Within the squadron, I was called 'the Professional Sadist'. I attacked anything—a bus on the road, a passenger train, even single cyclists."[34] The fact that a pilot is capable of saying something like that to another airman indicates a certain level of tolerance for such brutality.

The German fighter pilot Werner Mölders (see Chapter 3) was allegedly very careful to ensure that the airmen in his unit stuck to the "rules of war." It has been told that he on one occasion gave an airman a serious scolding in front of the entire unit for having fired on a civilian train in England with his machine guns. But Mölders seems to have been quite insensitive to the bombings of London. After a mission to escort bombers towards London, Mölders said contentedly, "The English have soon had enough. In London, there can't be a single whole window anymore!"

Now, this was nothing unique to the Luftwaffe. The South African fighter ace "Sailor" Malan also believed that it was better to damage rather than shoot down German bombers, "so that they com home to their comrades with dead and dying men onboard." In all of western Germany, the memory still lives on of how American fighter-bombers—"Jabos"—amused themselves with firing on anything that moved, even single farmers on workhorses.

There was also a kind of "code of honor" within large parts of the Luftwaffe. Heinz Rökker recounted, "I always only aimed at the engines and tried to avoid hitting the crew."[35] Heinz Rökker was a night fighter pilot, and this attitude was especially common among the night fighter pilots. Firing on downed pilots hanging in their parachutes did occur, but was often considered "disgraceful." Asked by Hermann Göring what he would do if any of his pilots fired at an enemy airman in a parachute, the fighter pilot Adolf Galland replied, "I would consider it murder and do everything in my might to have that person transferred from my group."

It seems as if it was more common for German airmen to be shot at while in parachute than them doing that to others. American fighter pilots in particular would fire on downed enemy airmen when they were hanging in their parachutes. But war hardens. One of the German airmen who was killed in this way was Ernst Süss, an ace with 68 aerial victories. Süss was very close friends with Alfred Grislawski, and he became a witness of the event. When I, somewhat naïvely, asked Grislawski forty years later what it felt like to have a comrade so brutally killed, he looked at me with a vacant face and said, "*Es war Krieg, Herr Bergström...*" (It was war, Mr. Bergström).[36]

What the airmen could experience when their comrades were killed was recounted by Lance Corporal Friedrich Rott from the attack wing III./ZG 2—here, in an account of how his wing commander, Captain Willi Hachfeld, died in connection with a landing in North Africa in December 1942, "Upon landing, his aircraft ran down a bomb crater and turned over and caught fire. He screamed like a wounded animal—it was horrible. You could hear his screams from among the flames in spite of some of our engines being warmed up. We couldn't stand it. It was impossible to help him because the ammunition in his aircraft was exploding. The mechanics gave the engines full throttle to drown out his screams."[37]

When an especially successful airman was killed, it could have a demoralising effect on the unit. But mostly, the airmen shook it off. Hans-Ekkehard Bob recounted, "When someone had been killed, he always had a bottle of liquor among his belongings. We shared it and toasted for him. Then we cleaned out his things. And then, we often didn't think any more about that. It was war."[38] The veteran aviators recount that the main cause for concern and stress was the bombings of the German home district and the growing difficulties for their families back home.

The fact that Jews were being persecuted was something everybody knew. Rumours of concentration camps probably reached most people. Many people had also learned about the mass murders and the death camps. The most common reaction was to put their heads in the sand, to disbelieve it or to block it out. Airmen such as Marseille and Mölders allegedly objected to the Nazi policy of genocide, but it has been impossible to get a definite confirmation of this. There was some antisemitism, but it does not seem to have been particularly tangible among the aviators. There were also pilots who were Jews and who were being protected by airmen such as Galland, Graf, Grislawski.[39]

It was quite common for airmen in the Luftwaffe to be convinced Nazis, but the opposite was also common. Lieutenant Antonius Wöffen from JG 27, secretly eavesdropped on during a conversation with a fellow prisoner in British captivity, was overheard saying, "Hitler is Germany's leader. But he's also digging Germany's grave."[40]

Fear of fighting was also present. Stress and exhaustion were more common among bomber pilots than among fighter pilots. According to the bomber pilots themselves, it was probably because the fighter pilots had a greater possibility to choose for themselves whether to enter or abstain from combat. In December 1941, the unit surgeons at KG 51 reported serious conditions of psychological exhaustions among the airmen: paroxysms of weeping, irascibility, and even epileptic seizures. Johann Maschel from bomber wing KG 2 recounted in March 1943, "We have an old observer who's still flying in the squadron. He's been flying since Poland in 1939. He's completely worn out. He's 23 but has lost all of his hair. He's completely emaciated. When you talk to him, he's so nervous that he stutters."[41]

The bomber pilot Hans-Georg Bätcher, who was 25 years old when the war broke out, recounted,

"The slightly older airmen were more frightened than the young ones, especially if they were married and had a family. But being cautious for the sake of your family wasn't always a good thing. Sometimes, they seemed to be daydreaming during a mission, and that was of course very dangerous."[42]

The Stuka pilot Erhard Jähnert admitted that "I was scared before every mission, every one of them,"[43] and the fighter pilot Heinz Lange said, "Anyone who says that he wasn't afraid, he's lying, perhaps to himself."[44]

The degree of fear and anxiety could vary considerably depending on the circumstances. Hans Philipp, who thought that "a skirmish with a Spitfire" could be "great fun," also said, "When you flew against the great American bomber formations, you reviewed every sin you had committed."[45]

But tolerance for fear was not always particularly great. Arthur Gärtner, a fighter pilot at JG 54, gives an example of this: "I heard one of the pilots scream on the radio: 'Help me! Help me! There's a Yak behind me! I was pissed off and shouted back over the radio: *Halt die Schnauze und kurbele weiter!*' [Shut up and keep turning!]"[46] Sometimes, the nervous breakdown came in the middle of aerial combat. You could see it in the enemy pilots, and this seems to have been a common phenomenon. Adolf Galland recounted, "You could be chasing an enemy pilot who was evading you cleverly for quite some time. But when he noticed that he couldn't shake me off, it was as though he just gave up. Suddenly, he stopped taking evasive action and just flew straight forward, so that I could shoot him down without any difficulty. Once, I flew past a Hurricane that I'd hit in that way, and saw that the pilot was sitting completely paralysed, his face in his hands."[47]

As strange as it might sound, the airmen were often more or less intoxicated during aerial combat. They simply had to take a schnapps or two ahead of each mission to keep their nerves under control. They used to call it *Zielwasser* (target water)—"in order to be able to shoot better," as they used to say. One of the veterans from the fighter wing JG 52 said many years after the war, concerning the ban on driving drunk in Germany, that if today's German authorities had been breathalysing the airmen during the war, the entire JG 52 could have been grounded.[48] "Assi" Hahn (see Chapter 2) recounted that "oxygen was perfect against a hangover."[49]

Apart from alcohol, the Luftwaffe airmen were often under the influence of methamphetamine—which the air force surgeons prescribed in large quantities, under the brand Pervitin, in order for the airmen to be able to stay awake. Quoting from Wikipedia's article on methamphetamine: "Methamphetamine (contracted from N-methylamphetamine) is a potent central nervous system (CNS) stimulant that is mainly used as a recreational drug… In low to moderate doses, methamphetamine can elevate mood, increase alertness, concentration and energy in fatigued individuals, reduce appetite, and promote weight loss… Chronic high-dose use can precipitate unpredictable and rapid mood swings, stimulant psychosis (e.g., paranoia, hallucinations, delirium, and delusions) and violent behavior… Methamphetamine is known to possess a high addiction liability (i.e., a high likelihood that long-term or high dose use will lead to compulsive drug use) and high dependence liability (i.e. a high likelihood that withdrawal symptoms will occur when methamphetamine use ceases)."

But alcohol, drugs and the "bubble" that the Luftwaffe airmen were living in did not stop them from critical reflections on life. Hannes Trautloft, commander of fighter wing JG 54, once heard Hans Philipp grimly philosophising about life, and he wrote it down in his diary, "I'm 25 years old. I've been awarded the Knight's Cross with Oak Leaves and Swords. I've shot down 150 enemy aircraft. I know everything about how to fly, how to aim off, how to kill your enemy. But I know nothing about life. I don't know how to lead a normal life, how to handle a job, how to raise children. I know the war, I'm a craftsman of death, but what will become of me after the war?"[50]

How, then, were the Luftwaffe veterans doing after the war? Today, we know that American veterans from Iraq and Swedish veterans from Afghanistan have been struggling with post-traumatic stress disorder (PTSD). But how was it with the Luftwaffe airmen? Did they feel any PTSD? When I have asked them, I have mostly received the kind of answer that, for instance, Alfred Grislawski gave me, "That hadn't been invented in those days."[51]

Last summer, I was happy to receive his son, Jürgen Grislawski, as a guest in my home. I then took the opportunity to ask him about this. Jürgen replied, "My father was feeling bad for ten years after the war."[52]

In the following chapters, we will go deeper into the realities for the Luftwaffe fighter pilots, both on the Western and Eastern Fronts, by taking a close look at a few individuals.

Luftwaffe Slang and Code Words

Börsenschluss
– (closing the stock market) radio silence
Caruso
– course
Dicke Autos
– (fat cars) four-engine bombers
Dödel
– (dick) The Knight's Cross
Dödel mit Blumenkohl
– (dick with cauliflower) The Knight's Cross with Oak Leaves
Dödel mit Essbesteck
– (dick with cutlery) The Knight's Cross with Swords
Eisenhaltig Luft
– (iron-bearing air) flak
Flakwalz
– (flak waltz) evasive manoeuvre from flak
Gartenzaun
– (garden fence) airport
Glasarschbegleitung
– (glass arse escort) escort for Fw 189
Goldfasanen
– (golden pheasants) party bigwigs in the Nazi Party
Indians
– enemy fighters
Katschmarek
– (a Polish first name) wingman
Klappertopp
– The rear gunner of an Il-2 Shturmovik
Lucie-Anton
– to land
Malinki partisan
– (Russian: little partisan) lice
Viktor
– Roger, affirmative (over the radio)
Yo-yo tactics
– To attack during a dive from above, dive underneath the enemy aircraft, the climb, fire at the enemy and continue climbing at high speed

The fighter general Adolf Galland (in white jacket) together with the commander of JG 54 "Grünherz", Hannes Trautloft (l.). On the other side of Galland is Dietrich Hrabak (see the chapter on "Assi" Hahn). (Photo: Galland.)

Two German pilot graves on the Eastern Front. (Photo: Trautloft.)

The future Luftwaffe commander Hermann Göring as a fighter pilot during the First World War. (Photo: Galland.)

The comradeship was everything. From left to right the German fighter pilots Heinrich Füllgrabe, Hermann Graf, Alfred Grislawski, and Ernst Süss. Shortly after this photo was taken, Grislawski saw his friend Ernst Süss being shot dead by an American fighter plane while hanging in his parachute. When I was quite naïvely asked Grislawski how he experienced it, he replied: "Es war Krieg, Herr Bergström ..." (It was war, Mr. Bergström). (Photo: Grislawski.)

The Air Force uniform, medals and celebrity made the pilots very popular with the opposite sex. Seen in this photo is Hermann Graf, who was the first to achieve 200 air victories, together with the famous film star Jola Jobst. (Photo: Wägenbaur/ Traditionsgeschwader JG 52.)

The Stuka pilot Theodor "Theo" Nordmann is received with flowers when he returns from his 600th combat mission on August 24, 1942. Nordmann conducted about 1,200 combat flights before he was killed in action on January 19, 1945. (Photo: Friedrich Lang.)

Hans-Ekkehard Bob (with camera) in front of his Messerschmitt 109 G-6 on the Eastern Front in 1944. Bob flew 700 combat missions between 1939 and 1945 and scored 60 air victories. After the war he became Europe's oldest active pilot. He passed away in 2013, at the age of 96. (Photo: Bob.)

Cohesion was particularly strong among the bomber crews, who cooperated very closely, often under intense pressure for many hours at a time. Here is the crew on a Heinkel 111. On the right is the pilot, next to him the observer and behind them the flight mechanic. (Photo: Bätcher.)

That the Luftwaffe airmen felt superior is evident from the facial expressions of both of these people—to the right the Luftwaffe pilot who steps ashore in Canada after being shot down and captured, to the left the British soldier who receives him. (Photo: via Galland.)

The remnants of a French refugee column after a strafing attack by German aircraft 1940. (Photo: Hahn.)

Erich "Bubi" Hartmann, the most successful fighter ever, with 352 air victories, in front of his Messerschmitt 109 in the spring of 1945. The red heart was the squadron badge of 9./JG 52. Erich Hartmann passed away in 1993, 71 years old. (Photo: Hartmann.)

The fighter pilot Hans-Joachim Kroschinski poses happily in front of his Messerschmitt 109 while this is being serviced. Kroschinski conducted 360 combat missions, including 240 low-level attacks against ground targets, and was credited with 76 air victories. He died in 1995, at the age of 75. (Photo: Kroschinski.)

These two Messerschmitt 109 Fs collided during a landing on the Eastern Front in 1941. (Photo: Leykauf.)

Two of the opponents: Soviet fighter pilot Aleksandr Pokryshkin (left) and Dmitriy Glinka in front of one of their Airacobra fighter planes. Pokrysjhkin flew 650 combat missions and participated in 140 air combats. His official score is 59 air victories, but recent research by Russian aviation historian Michail Bykov shows that the actual number was 46, as well as six shot down in collaboration with others. Dmitriy Glinka flew 300 combat missions and was credited with 50 air victories in 100 air combats.
(Photo: Kovatchevich)

Air victories and other qualifications were carefully recorded. Here Hans Philipp proudly poses in front of the rudder of his Messerschmitt 109. Each bar symbolizes a shot down enemy aircraft. "Fips" Philipp had come up with 206 air victories when he himself was shot down and was killed during an action against US four-engine bombers on October 8, 1943. (Photo: Trautloft.)

On November 15, 1943, German aviator Heinrich Bartels claimed to have shot down four U.S. fighter planes in two minutes, which allowed him to paint his 70th bar on the rudder of his Me 109. Bartels was shot down and killed on December 23 1944, shortly after scoring his 99th victory. He became 26 years old. (Photo: Neumann.)

Heinkel 111 bombers heading towards enemy territory. (Photo: Vollmer.)

CHAPTER 2

"Assi" Hahn:
The RAF's Nemesis who met Stalin's Son

Of all the German veteran pilots whom I have met, none was probably as funny and good-natured as "Assi" Hahn. In his first letter to me—in 1973, when I was only 15 years old—he enclosed a photograph where he is standing next to a donkey. Apparently, he felt compelled to explain in the letter: "The one with a hat on is me. The other is a donkey." A few years later, I visited him in his home in picturesque Goslar.

As we were sitting in his living room with his cognac glass and my cup of tea on the table with a map over the English Channel coastal area during the Battle of Britain in mosaic, he made the hours fly with his cheerful art of storytelling. He had a good-hearted and roaring laughter and kept seasoning his stories with the odd joke or humorous comment. It was quite obvious that he was very intelligent. His great social skills made me forget that I was just a young lad visiting the famous veteran pilot in his home.

Only when we veered into his time in Soviet captivity did he become serious; then, I saw how as if a dark shadow crossed his face. It was obvious that his almost seven years in Soviet captivity, longer than most German soldiers, had affected him very badly. In fact, his time in captivity could have become little more than a few days, ending with a humiliating and brutal sudden death. If it had not been for the man who was commander for Joseph Stalin's son Vasiliy. But neither of us knew anything about that by then. Not too long afterwards, just before Christmas 1982, Hahn passed away. The good life was too taxing on him.

Because "Assi" Hahn had been a true *bon vivant* already during the war, that has been confirmed by many other veteran pilots. He enjoyed all the good that came with life, and seems to have lived the part of a future knight at a manor somewhere in the East—which was what all who had been awarded the Knight's Cross had been promised by Hitler after the war, when the "final victory" had been won.

One of "Assi's" comrades recounted with a certain amount of distaste that "Assi" had once been talking about this, jerking his hand as when you wield a whip, saying "then the serfs will know their place." To what extent Hahn actually meant this is difficult to say, but it would probably also be typical of him to blurt out such a thing as a kind of a joke in convivial company. After all, he was only in his twenties when he was supposed to have dropped that remark, and you need to remember under which circumstances he did it. The fact that he also enjoyed a drink or two surely also had something to do with it.

I have been talking to many of Hahn's old comrade airmen, and with one single exception—the one who told me the above—they describe him as a very kind and straight-forward type of person. He had incredible authority, and one of his comrades said that "his superiors were afraid of him." That could be an exaggeration, but he was clearly very popular among his subordinates, who all describe him as a very understanding officer and a good commander. He also enjoyed socializing with downed and captured enemy airmen—a tradition that harked back to World War I.

As you read his letters and reports, you get the impression of a person with quite a portion of self-confidence, and you could probably say that he was not afraid of drawing attention to his own strong sides. Which almost cost him his life more than once. More on this later.

This chapter is mainly based on interviews with "Assi" Hahn, carefully corroborated with wartime documents.

"Assi" was born in Gotha on April 14, 1914, and as he grew up, he was torn between his enthusiasm for sailplane gliding and track and field. He turned out to be as talented in both areas. At age twenty, he volunteered for military service—this was before Hitler had brought back conscription—and was one of the first ones to join the new Luftwaffe when this was formally formed in early 1935.

The following year appeared to become his greatest ever, when the 22-year-old Hans Hahn was selected to represent Germany in modern pentathlon at the Berlin Olympics in August 1936. However, disease stopped his participation at the last moment! It must have been with mixed feelings that Hahn watched as another Luftwaffe pilot, Gotthard Handrick, won the gold medal with 31.5 points ahead of the American Charles Leonard—whose brother Bill, by the way, also became a fighter pilot during World War II.

But instead, his career as a fighter pilot now took off seriously. In the spring of 1936, he was promoted to Second Lieutenant, and was stationed at a fighter group where he learned several advanced tricks in the air from the latterly so famous First Lieutenant Werner Mölders. He would soon discover what a talented aviator the young Hahn was, and in November 1937, Hahn was appointed a flight instructor himself at the new fighter pilots' academy at the brand new airbase at Werneuchen a few dozen kilometers northeast of Berlin. As an officer and flight instructor, Hahn was invited to the inaugural ceremony for the airfield and the unit at the Werneuchen town hall—where he and the other aviators each received presents consisting of a painting by Hitler and a copy of *Mein Kampf*...

Here, Hahn became good friends with the unit commander, the fighter ace from World War I, "Theo" Osterkamp, and his young wife. But training newly-graduated pilots did not suit the young Hahn's fiery temperament—he wanted to see action, and was hoping that the conflict with Czechoslovakia over the Sudetenland would give him the chance to prove himself. Now, this was no unusual attitude among the young hotspurs of those days, wearing Luftwaffe wings on the chests of their uniforms. In any case, "Uncle Theo" Osterkamp believed that Hahn would be more useful at a regular fighting unit.

Osterkamp arranged for him to be prematurely promoted to First Lieutenant and transferred to a fighter unit equipped with the new, revolutionary fighter Messerschmitt 109. By then, in February 1939,

Hahn had already proven himself behind the controls of a '109 in Werneuchen.

Hahn was quite lucky. In his first fighter unit, he had learned quite a lot from Werner Mölders. The group commander he had in his new fighter unit, JG 231, was Captain Johannes Gentzen, who was a very skilled aviator himself. Gentzen was one of the chosen few who had been receiving secret flight training in 1932 at the Soviet airbase in Lipetsk. It was he who taught Hahn how to remain calm during aerial combat and to wait for the right moment for a surprise attack rather than attacking immediately.

The outbreak of war in September 1939 was followed by disappointment for Hahn. The unit that he belonged to was stationed in mid-Germany, far from any frontlines, and at that time, there were no enemy aircraft flying in over Germany to bomb its cities. While hearing and reading the reports about his former division commander Gentzen—who was raking home one aerial victory after another until he was the No. 1 top ace of the Luftwaffe[53]—he himself would only fly boring routine missions over Germany.

Late in the fall of 1939, Hahn and yet another few pilots from the Zerbst wing were suddenly stationed as the first pilots in a completely new fighter wing, the second group of Fighter Wing No. 2 (II./JG 2). This wing carried the honorary name "Richthofen" after the famous "Red Baron" from World War I, and as new group commander, a major celebrity arrived, Captain Wolfgang Schellmann. He still retained a light tan from his time as a volunteer pilot during the Spanish Civil War, where he, with 12 victories, had become the second most successful fighter pilot of the "Condor Legion," after Werner Mölders. Hahn was appointed commander of the 4th squadron of Fighter Wing No. 2 (4./JG 2).

However, neither did the transfer to the "Richthofen" wing mean any direct participation in combat. Once again, Hahn found himself far away from the heat of the battle. He had missed the Spanish Civil War, he had missed the Polish Campaign and the aerial battles over the Western border, which yet remained calm. In early April 1940, the unit was then suddenly stationed at Neumünster near Kiel, and the rumor was than now—now there would be fighting. Two days after the arrival to this new airbase, the news came: Germany had invaded Denmark and Norway.

But there was still no fighting for Hahn and his comrades. When the next deployment order came on April 12, it was not to fly north to Norway, but instead, the journey went back to Nordholz in northwestern Germany.

In the evening of May 9, 1940, Schellmann summoned his men to an important meeting. Hahn's heart leaped when he heard that the offensive in the West would break out the next day! On the day of the attack itself, there was frantic activity. The unit would be deployed at an airfield closer to the Dutch border. They arrived there on May 11, and the very next day, Hahn and the other pilots escorted German bombers against Dutch targets. He was able to spot the great clouds of smoke from the completely burned-out downtown Rotterdam—the result of a Luftwaffe raid that was actually supposed to have been cancelled—before the unit at dawn on May 14 was transferred to a captured airfield in northern Belgium. From there, Hahn took off almost immediately, leading his squadron on a mission over the combat area.

The weather was hazy and visibility was miserable, so they did not see much of the fighting below them, other than weak flashes of light flaring up on the ground every now and then.

Captain John L. Sullivan was one of Canada's famous Mounted Police who had trained as a pilot. When World War II broke out, he volunteered for frontline duty and was placed at the RAF's all-Canadian 242 Squadron, in England. But for the British air units in France, the sudden German attack meant a shockwave of losses. One of the fighter units, No. 607 Squadron, lost five of its twelve Hurricanes and three pilots during the first three days. A quick order went to the Canadian fighter squadron to transfer four aircraft and pilots to 607 Squadron. Sullivan, Willie McKnight (a future ace during the Battle of Britain) and two other pilots set off on the 13th. Arriving at the airfield at Vitry-en-Artois late in the afternoon, they met a pitiful state of things. Another three Hurricanes and one pilot had been lost in combat with the Luftwaffe's Messerschmitt 109s that morning. The entire front seemed to be dissolving and the mood was at rock bottom. But the arrival of the self-confident Canadians gave the survivors new hope.

In the morning of May 14, Captain John L. Sullivan took off leading the Hurricane planes and flew against the German units that were on their way through France. It was just before ten in the morning, local time, when they spotted a formation of German Henschel 123-type fighter-bomber biplanes through the haze. The attack came as suddenly as unexpectedly for the German aviators. Two Henschel planes were shot down immediately.

"Hurricanes!" someone cried over the German flight radio, but Hahn had already seen them. The Germans had altitude superiority and the British did not seem to have discovered them. Hahn recounted that he "only did as I had been taught": he swept down and went into position behind the nearest Hurricane. Mölders and Schellmann had told him to keep cool and wait with firing until he was really close. He must have been as close as only ten meters from the British aircraft—whose large tailfin was towering mightily in front of the turning propeller of the Messerschmitt—when he pressed the fire button as hard as he could. Hahn was shocked by the sudden bank, the flames that flared up, and the debris from the Hurricane that came swirling around his own aircraft.

John L. Sullivan could hardly have reacted at all. He must have been killed immediately when the 20-mm shells and machine gun bullets tore his cockpit to shreds—this was before the British had learned to fit rear armor behind the pilots; they believed that the downside of the increased weight outweighed the advantages.

More out of fear of all the debris from the Hurricane dancing around in the air than due to any calculation, Hahn pulled the control column towards him—and immediately had another Hurricane in front of him. He fired again, only seconds after the first burst of fire, and this enemy machine also went down in flames.

When Hahn came around again after turning, all British aircraft were gone as if by a stroke of magic. Did we shoot them all down? he was able to think before the radio exploded with cheers in German. "Shut up! We did great!" Hahn cried back—radio discipline was not the very best.

But when the Messerschmitt airmen returned to base, they could not control themselves any longer. They swept across the field with roaring engines, rocking their wings as a sign of success. When Hahn had landed, the ground crew flocked around his aircraft. He was lifted down onto the ground and was carried in triumph for a full lap of honor around the apron.

His two kills were the first ones for the unit. In addition, a young sergeant had shot down a third Hurricane. Much later, it would turn out that Hahn's first, fast victory in that battle was never confirmed by the bureaucrats at the Ministry of Aviation. Therefore, his second claimed downing was officially confirmed afterwards as his first aerial victory.

Exaggerating the number of victories in aerial combat at lightning speed with one's body boiling with adrenalin was rather the rule than the exception during the aerial battles of World War II. At the Ministry of Aviation, there was an entire division with no other task than trying to analyze the pilots' claims to find out how high losses had actually been inflicted on the enemy. No other country had such rigorous reviewing of claimed downings. Of course, they could not assess whether or not the pilots had been intentionally laying it on a little thick—something that was certainly not unusual—but they tried their best.

In this case, however, you can quite clearly see that Hahn and his wingman were not guilty of any "fishing stories" when writing their reports—on three different sheets of paper—about each of Hahn's downings. At the National Archives in Kew in southwest London, 607 Squadron's report from this aerial battle has been preserved. There, it says that Flight Lieutenant John L. Sullivan was the first one to be shot down, while he was busy attacking a Henschel 123 himself. Immediately afterwards, Flying Officer Gerald I. Cuthbert and Pilot Officer

Arthur Le Breuilly were shot down. Comparing this with the German reports shows that Sullivan was unambiguously shot down by Hahn—whose claimed downing was thus absolutely correct, even though the bureaucrats in Berlin drew another conclusion. Cuthbert crashed his damaged aircraft at Aische-en-Refail, a few dozen kilometers away, and should therefore be the one shot down by the other German pilot. Le Breuilly died as his Hurricane ("AF-K") was buried in the ground just next to Sullivan's crashed machine. All Hurricane pilots were killed. In Hahn's squadron, no aircraft was even damaged.

607 Squadron's death march continued. The next day, another three Hurricanes were lost. With one of them—which was shot down by the German ace Mölders—the unit commander, Squadron Leader Lance Smith was lost. It was the second time in four days that he was shot down, but this time, he was killed. On May 20, Flying Officer "Bobby" Pumphrey and Pilot Officer Richard Demetriadi were shot down in the unit's last two Hurricanes—including the four that had been brought there by the Canadians. The machine that Pumphrey went down in, "AF-H," had been damaged and repaired twice before, on May 11 and 13.

However, the men at II./JG 2 noticed on May 19 that the British could hit back, when two of their comrades were shot down by Hurricanes. In return, they shot down three of the British fighters. With one of them, Hahn finally had his second confirmed aerial victory. During another mission that day, he claimed to have shot down a French Morane 406 fighter. On that day, giant aerial battles were raging over the battlefield. The German fighter pilots claimed to have shot down 61 enemy planes against 20 own aircraft lost, while the true Allied fighter losses were 42 destroyed and 15 damaged aircraft.[54] In this confusion, it is impossible to determine whether or not Hahn's fourth claim was consistent with a real Allied loss.

Neither was this downing confirmed by the Ministry in Berlin, but the airmen at II./JG 2 knew nothing about this yet. As for now, Hahn was considered the most successful pilot from the wing in the war, with four aerial victories, which were more than even Schellmann's three. Schellmann, however, would catch up. When he could count his sixth aerial victory (fourth confirmed) near Paris on June 3, Schellmann pulled off his seventh.

The kind of competition for downings that went on between Schellmann and Hahn was something that was encouraged by the Luftwaffe commander Göring, and that also spurred the young fighter pilots into greater aggressivity and zeal during aerial combat. On June 6, Hahn saw an American-made French Curtiss Hawk 75-type fighter coming down right in front of his weapons. This was in the middle of Operation "Paula," the last major battle between the Luftwaffe and the French air force as the Germans made an effort to eradicate its adversary's air force. In an extensive aerial battle, the Germans claimed to have shot down 85 enemy aircraft, against 19 own losses in combat. With so many aircraft participating, exaggerations—or, rather, double counting—were inevitable. Of the French losses, 17 were fighters with radial engines (Hawk-75 and Bloch 151/152), while the Germans reported shooting down 22 Blochs and 16 Hawk-75s.[55] When France capitulated a little more than two weeks later, Hahn was equal to Schellmann with seven aerial victories as the best pilot of the wing.

Not only on the German side did most people now believe that the war was more or less over—the badly battered British on their island would probably soon throw in the towel. Hahn, who was now dubbed "Assi"—The Ace—was already somewhat of a "star," and could take certain liberties. One of these was to get an entire little mini zoo of his own. He had already "deployed" his famous donkey, "Icko," at the unit. Now, it was followed by a large eagle owl, which he christened "Moritz" after the cartoon figures "Max and Moritz," a parrot, and a huge male Great Dane dog which he named "Lux." Later, as "Lux" had grown, he used to run after his master's Messerschmitt 109 as it rolled down the runway after a sortie, and—to the horror of the mechanics—leap up onto the wing as soon as he caught up with the machine. What drew the largest attention by far was the bear cub that the Berlin Zoo donated to the unit!

But, as we know, the war was far from over. The Battle of Britain, which now followed, would become something completely different for the German aviators, who had gotten used to victory. The pilots of the "Richthofen" fighter wing were given perhaps the most difficult tasks on the German side: they were based in the Normandy area, which meant that they had a much longer approach over the English Channel before reaching the British mainland than the air units at the Pas de Calais area had. This meant that the British fighter units were often given good advance warning by radar, and were able to stay at a high altitude and wat for the Germans as they came gliding in over the coast.

On August 11, 1940 the entire "Richthofen Wing"—a total of 130 Messerschmitt 109s—were made ready to participate in a major operation aimed at the naval bases at Portland and Weymouth with the main objective to lure the British fighter command into battle. The first ones to go were the bombers together with the twin-engine Messerschmitt 110 fighters. These clashed with 70 British fighters in an aerial battle that cost both sides dearly.

The "Richthofen Wing" and fighter wing JG 53 had been tasked with following up and relieving the Messerschmitt 110s, but that did not turn out well. In spite of the battle starting to wane, and many of the British fighters having been shot down or having used up their ammunition, the encounter with the British fighters cost the "Richthofen Wing" a loss of eight Messerschmitt 109s. In Schellmann's group, two pilots were lost, while the Spitfire that "Assi" Hahn shot down was the only success for the unit.

As the month of August segued into September, three weeks of aerial combat had cost II./JG 2 one third of its aircraft supply and nine out of thirty-nine pilots. The survivors could calculate their chances of survival on the fingers of one hand. Against this, they had only 13 own downings. Five of these were due to "Assi" Hahn.

The situation improved somewhat as the bomber units in the Normandy area switched to night raids and the fighter command there was regrouped to Pas de Calais. On August 31, "Assi" flew two missions over southwest England and was credited with the downing of three Spitfires—the only success for the wing that day.

The British fighter command was a terrible adversary in the air over southern England. "Assi" Hahn was the commander of the wing's 4th Squadron. His colleague, the commander of the 6th Squadron, First Lieutenant Edgar Rempel, was one of those who were downed and killed during the major operation on August 11. His successor, First Lieutenant Karl Müller, was killed in action over the Strait of Dover on September 4.

Had it not been for "Assi" Hahn, who kept morale and moods up as best as he could, the fighter group could very well have become yet another one in the line of Luftwaffe units that had to be taken out of combat after morale collapsed. Hahn was now the group's leading ace. For his great success and his ability to keep the spirits up in these difficult times, he was awarded the Knight's Cross on September 24, 1940. By then, he had reached 20 victories. Soon afterwards, he was promoted to Captain.

On October 19, there was a sudden order to the unit to immediately fly to a base in southwest France for a temporary but top secret mission. As it turned out, the pilots were to protect a special train near the Spanish border. The most important passenger on this train was the Nazi leader himself. On October 23, he met with the Spanish dictator Franco in the small French border village of Hendaye, while "Assi" Hahn and other German fighter pilots were circling in the air above, keeping watch.

When the unit returned a few days later to the Channel area, a new surprise was awaiting "Assi": he was appointed commander of the third group of the "Richthofen Wing," III./JG 2. This unit was grappling with deteriorating morale among the pilots, and there was actually only one of them with any downing results worth mentioning: Master Sergeant Werner Machold, who, with 26 aerial victories—even more than Hahn—was the most successful pilot of the entire group at that time. The previous wing commander was sent to Germany to become a flight trainer instead, officially to spare his parents, who had already lost their two other sons in the war.

Under "Assi" Hahn's command, III./JG 2 would develop from being the least successful group of the "Richthofen Wing" into becoming the most successful one. When he took over, the Ist Group was credited with 259, the II. Group with 122, and the III. Group with 76 aerial victories. Until the turn of the year 1941/1942, the Ist Group pulled off yet another 73 victories, the IInd Group 144, and the IIIrd Group 69.[56]

It was "Assi" Hahn's inspiring leadership that caused the young Lieutenant Egon Mayer to really boost his results in air combat; when he was killed in action in March 1944, he was one of the highest decorated soldiers in Germany, with the Knight's Cross with Oak Leaves and Swords and 102 aerial victories on his tally. When Hahn assumed command of III./JG 2, Mayer had been in the "Richthofen Wing" for as long as Hahn had, and had flown scores of sorties, but had only with great difficulty managed to have the downings of two enemy aircraft credited to him—one French Morane 406 in June 1940 and one Hurricane in October. As if this were not enough—Allied loss reports show that the first one of these, if not even both of them, did not correspond to any real losses.

The winter of 1940/1941 was calm for the pilots at III./JG 2, who were stationed in quite a peaceful corner of France. The Battle of Britain had ended with British victory and both sides were now waiting. "Assi" used the time to encourage his pilots anew.

In June 1941, it all broke out again—ironically, at the same time as Hitler showed that he had finally given up the attempts to defeat the RAF and moved almost all of his air force eastwards to attach the Soviet Union; Operation "Barbarossa" was begun on June 22, 1941. The same day, III./JG 2 was transferred from the calm Normandy area to Pas de Calais—arriving just as the British air force deployed its "Inverted Battle of Britain." Now, the British air force was on the offensive, attacking the German airbases in northern France, and the Germans were on the defensive.

The first few days were a terrible lesson for the German fighter pilots, who were completely unprepared for meeting such great formations of Spitfires over their own territory. Between June 22 and 25, eight pilots from III./JG 2 were shot down and killed. But when the Germans had recovered, they adapted their tactics to the new situation. As the British came in over Calais with their bombers on July 7, it was the Germans that were waiting for them. They cracked down from above and drove away the Spitfire pilots. When the Messerschmitt pilots had landed again, they counted a total of six kills—two of them by "Assi" Hahn—against only one loss of their own. The next day, seven Spitfire pilots did not return from a raid over France while the Germans lost three Messerschmitts. Once again, Hahn was successful, with one aerial victory.

"Assi" Hahn and his pilots now flew then latest version of Messerschmitt 109, the F version, which was both faster and more agile than the previous E version, which they had been using during the Battle of Britain. With this aircraft, they gave their British opponents a severe beating on July 10, when 190 Spitfire planes were escorting three four-engine Stirling bombers against a German airfield. Without losing a single aircraft of their own, the Germans shot down eight Spitfires and one Stirling. As for Hahn, he reached his 30th victory with two new downings, and Egon Mayer was credited with his 13th victory—nine of which were since June 24.

The result for III./JG 2 during July 1941 was 57 British aircraft shot down—all of them Spitfires—against four own losses.[57] On August 14, "Assi" was awarded the Oak Leaves to the Knight's Cross, which only 31 soldiers in the German Wehrmacht had received previously.

By the end of the month, he had 41 aerial victories, while his "protégé" Egon Mayer had reached 21. He, too, could now also boast a Knight's Cross—which, in spite of all that has been said, did mean a lot to most German soldiers (and still does to many of the surviving veterans).

It was also, in much, thanks to "Assi" Hahn's leadership of his unit that the British air offensive against the German airbases in France ended in fiasco. By this time, there were virtually only two German fighter wings in the West, JG 2 "Richthofen" and Adolf Galland's JG 26 "Schlageter." Between June and December 1941, the air war in the West cost the British fighter command a loss of 411 fighters, mainly Spitfires, while the Germans lost 103 fighters.

With each month, the Germans learned more and more, and the result of the aerial battles became increasingly uneven in their favor, in spite of them being inferior in numbers. During the last three months of that year, III./JG 2 was credited with the downing of no less than 63 British fighters against one single own loss in combat! That was a better victory to loss ratio than at any other German fighter unit, including the Eastern Front, during this period.

Among the German veteran airmen, it is quite well-known that pilots at JG 2 "Richthofen" were guilty of large exaggerations concerning their own successes; it is not unusual that the "Luftwaffe gossip" alleges that the "Richthofen airmen" developed a culture where deliberate exaggerations were accepted in a completely different way than at many other units. Comparisons with Allied losses show that that could actually have been the case. But that is something that occurred later during the war (with the exception of one certain group—not the one Hahn belonged to—during the Battle of Britain). On September 20, 1941, JG 2 claimed to have shot down twelve Spitfires, three of them by Hahn. The true British losses were eight Spitfires.

Egon Mayer began the new years with shooting down the Swedish volunteer Ralf Häggberg, who flew a twin-engine Westland Whirlwind fighter with the RAF. This was on February 12, 1942, when the "Richthofen Group" was escorting the German warships *Scharnhorst* and *Gneisenau* straight through the English Channel—an operation that was as great a success for the Germans as it was a humiliation for the British. "Assi" Hahn himself contributed with shooting down a Spitfire as his 50th aerial victory.

The year 1942 was no better for the British fighter command than 1941 had been, and the German fighter pilots at the English Channel were gaining successes that they had never experienced before. The pilots of JG 2 and JG 26 claimed 972 aerial victories during that year—corresponding to about 900 real British fighter losses and a smaller number of bombers (in daylight). Meanwhile, JG 2 and JG 26 lost 113 aircraft in combat.

This ratio—eight British fighters lost for every downed German fighter—has several explanations. To begin with, it was because of the tactics that the British applied—they came marching in over the French coast in large, ungainly formations, while the Germans were lying waiting for them, and then attacked them with smaller formations of fighters in rapid attacks from above. In addition, the British had the rule that their pilots were taken out of frontline service after a certain number of sorties to train new airmen. The Germans, on the other hand, kept their airmen in frontline service, so that they could accumulate combat experience that none of their British adversaries reached. The third factor is that the Germans had developed a new fighter, the radial-engine Focke-Wulf 190, which was clearly superior to the Spitfire models of those days.

The Focke-Wulf Fw 190 had been put into service at the English Channel in the fall of 1941. In May, 1942, it was time for "Assi" Hahn's III./JG 2 to replace their, by now, quite old Me 109 Fs with Fw 190s. After one month's training on the new aircraft, it received its baptism of fire with "Assi" Hahn at the controls on June 6, 1942. In two minutes, he shot down three Spitfires—his aerial victories Nos. 63-65.

By the late summer of 1942, he was at home in Germany on leave, and missed the greatest aerial battle over the English Channel since the Battle of Britain—the one that took place in connection with the failed Canadian raid at Dieppe on August 19. But his airmen were coping well without their popular commander. JG 2 and JG 26 claimed to have shot down 98 enemy aircraft against 23 own aircraft losses. The true British losses amounted to 106. III./JG 2 was credited with 13 aerial victories against one single own loss.

When "Assi" was back at the "Channel front" again, the British had begun deploying large numbers of the new Spitfire version, Mark IX, which meant that they had a fighter that, for the first time, could match the Focke-Wulf 190. On one of his first sorties upon his return, on September 16, Hahn shot down his first Spitfire IX.

Ten days later, ten Fw 190 pilots from III./JG 2 clashed for the first time with an American fighter unit, 4th U.S. Fighter Group, which was equipped with Spitfire IXs. In a drawn-out, swirling aerial battle that went on for almost twenty minutes, one American pilot after another was shot down. Not a single one escaped. Eleven Americans were shot down—four were killed, six were captured, and one got away and made it to Spain with the help of the Resistance.[58] The German losses were limited to one 190, but the pilot made it. The only American who made it back to base from that mission was Second Lieutenant Don Gentile, who had aborted prematurely because of engine failure and flown back to base before the Germans appeared. (Later, Gentile would develop into one of the most successful fighter aces in the USA during the war, with 21 aerial victories credited to him.)

"Assi's" downing on September 16, 1942, his 66th in total, turned out to be the last one at the "Richthofen Group." In October, 1942, there was a surprise order that he was to report to Major Hannes Trautloft, commander of JG 54 "Grünherz" on the Eastern Front.

It was with the greatest reluctance that "Assi" left his men at the "Richthofen Wing" and the comfortable life in France to go to the Russian winter. Initially, things did not look that bad. He had expected 30-40 degrees C below zero. But when he arrived at the Relbitsy airfield near Lake Ilmen south of Leningrad, there was a thaw, and the temperature was around 0 C.

"Assi" Hahn's usual old optimism came back to life. He had heard about the absolutely astronomical figures of shot down enemy aircraft on the Eastern Front, and, filled as he was by the Nazi propaganda, he expected quite a simple time where there would be ample opportunities for great successes in aerial combat. But there, he would have to think again. He soon discovered that he had gotten the completely wrong end of the stick concerning most things about the war in the air on the Eastern Front.

On November 22, 1942, he reported to Trautloft, who was an old acquaintance from the fighter pilot school in Werneuchen. Trautloft told him that he had had to give up a very successful commander of the second group, II./JG 54, as he had been appointed wing commander at JG 52, and that he had requested a successor with three qualities: he needed to be an experienced and confirmed successful and popular unit commander, a skilled fighter pilot and someone with experience of the Messerschmitt 109. "You, my dear Hahn, fulfill all these criteria," Trautloft said, and told him that when the fighter commander in the East, Lieutenant Colonel Günther Lützow, had proposed "Assi" Hahn, he had been very satisfied.[59]

Under the previous commander, Dietrich Hrabak, II./JG 54 had been formed into a well-managed and very successful fighter group with morale that was sky-high. Hahn was soon able to see that it was a sheer elite unit; its pilots were decidedly more experienced than those at III./JG 2. Two of his new subordinate airmen stood out especially— the 30-year-old Austrian Max Stotz and Hans Beisswenger. Both had been fighter pilots since before the war, both wore, as did Hahn, the Oak Leaves to the Knight's Cross (which not even Trautloft did), and both had just recently won their 100th aerial victories. Hahn was especially impressed by Captain Heinrich Jung's 4th Squadron, which did not have a single pilot without aerial victories of his own.

Hahn did not make his new wing commander disappointed either. On November 26, Trautloft wrote in his diary, "I took my Me 108 and flew to Relbitsy to formally hand over command of the IInd Group to Captain Hahn. It turns out that 'Assi' Hahn has already made himself at home there and begun 'ruling'. I'm sure he is the right man for the IInd Group."[60]

Neither did he make his new subordinates disappointed. This wing flew the latest type of Messerschmitt 109s, G, "Gustav." This had been equipped with a more powerful engine and stronger armament. Hahn felt very comfortable with his new machine and impressed even the aces Stotz and Beisswenger with some daring aerobatics over Relbitsy. Het got on especially well with Max Stotz; it turned out that both of them had the same sense of humor and talent for practical jokes and pranks. Eventually, they would begin flying as flight leader and wingman.

A few days later, the period of relative calm that had prevailed in this sector of the Eastern Front was broken as the Red Army attacked south of Lake Ilmen. On December 4, 1942, the alarm went at Relbitsy—a large Russian formation of aircraft is reported as incoming! "Scramble!" The whole group took off as soon as possible. Hahn ordered them to gain altitude as rapidly as possible. As the Soviet aircraft appeared from out of the have over in the East—Il-2 Shturmovik fighter-bombers at 500 meters' altitude with their fighter escort at about 1,000 meters—"Assi" and his pilots were higher up. They cracked down on the Russians in the same way as Hahn so many times had dived down on British Spitfires. But the fighters Hahn went after skillfully avoided his fire. "Assi" had not expected anything like that, but pursued the Russian, who now dove to escape. "Watch now how Daddy's teaching him," he called out over the radio. Just above the snow-covered ground, the Russian recovered his plane from the dive. Hahn fired, as he had so many times before, with a lead. The Russian flew straight into his fire and slammed into the ground, flames spurting out of his aircraft.

"Assi" pulled back the stick and turned while climbing to get to the Il-2s, which were steering straight towards the airbase. He soon called up with them and opened fire on the nearest one. To his surprise, he was able to conclude that neither this Russian nor the other assault planes made any evasive maneuvers, but continued straight towards the airbase. The Il-2 was heavily armored, but the armor-piercing shells from the '109's three automatic cannon plowed straight through the armor. The rear gunner vanished in a cauldron of explosions. Still, the Russian continued flying as if nothing had happened. "Assi" reduced the throttle so as not to fly past the slow-flying Ilyushin, aimed once more, and fired again. He fired short salvos. Eventually, the Il-2 started letting out a small stream of smoke and went into a slight dive.

Right then, tracers came flying up from the ground. "Assi" made way. Right in front of him was his own airfield, and it was from there that the flak was firing—against the Il-2s that had now started approaching against the target.

"Assi" was shaken as he landed. Luckily enough, the Russians had only carried out a quick attack and then departed again, otherwise, the damages could have been worse. But a few aircraft had been damaged on the ground, and several men from the ground staff had been wounded by shrapnel from the bombs that had been dropped. "Assi" had not experienced anything like that at the "Channel Front." There, he had seen the British urgently drop their bombs as the German fighters were approaching them, and them actually causing any damage to the German airfields was unusual.

The Il-2 that Hahn had fired on was seen crashing some distance from the airfield. Apart from "Assi's" two kills, his pilots reported the shooting down of three Soviet fighters and two Il-2s. In return, one of the Me 109s was lost.[61]

But the day was not over. Just before lunch, the alarm went off again. New Russian aircraft incoming! This time, Beisswenger and Stotz and their two wingmen took on the fighter escort, while "Assi" and the rest of the group positioned themselves behind the Il-2s. As the Soviet assault pilots noticed that their fighter escort had disappeared, they went down to the deck. But they stood no chance against the '109s. The first Il-2 was felled by Hahn's first salvo. What followed then was sheer massacre. When it was all over, seven Il-2s had been shot down. Two of them were on Hahn's account.

During the next few days, there was violent fighting in the air of the same kind—the Germans attacked formations of Il-2s relentlessly attacking targets in the German rear. As over the English Channel, Hahn had much success. On December 30, both sides made a maximum flight effort. In two air combats, he shot down four Soviet fighters—his best result for one day so far. Soon afterwards, he was promoted to major.

Trautloft was increasingly convinced that he had had the best possible successor to Hrabak. He wrote in his diary, "Major Hahn, this wing commander so full of vitality, distinguishes himself for his jokes, his sense of humor, and his good spirits. Moreover, he has an absolutely incredible memory and surprises us all by being able to recite almost all stanzas from the Niebelungenlied by heart. He holds a firm grip over the group and leads it from one success to another. He is a fantastic aviator himself and is very successful in aerial combat." [62]

Deep inside, however, Hahn had misgivings. He had decided to create an "expert group" (*Expertenschwarm*) consisting of him as its leader, Stotz and Beisswenger, and a promising young pilot named Albin Wolf. It was together with these that he had been able to achieve his best results for one day on December 30.

On January 12, 1943, Hahn had an urgent phone call in Relbitsy. It was Trautloft. The Russians had launched a major attack east of Leningrad and broken through—Hahn immediately had to regroup to that area![63] After a day with weather bad enough to make it impossible to fly, Hahn took off together with Stotz and flew towards Schlüsselburg on the southern shore of the Ladoga. What happened next would become "Assi's" most successful aerial battle ever. The two aces swept down on a formation of Lavochkin La-5 fighters, shot each of them in flames, climbed at excess speed, turned around and swept down on them again, with the same results. These Soviet pilots turned out to be very inexperienced. In three minutes, the two Germans carried out as many lightning attacks from above, without the Russians actually offering any resistance. Stotz was credited with four and Hahn with three victories.

Soviet sources confirm the image, even if Stotz and Hahn overestimated both the number of La-5s as well as their own downings. The Soviet unit was the 263rd Fighter Regiment, and an entire four-plane group was shot down by the sudden attack: Second Lieutenant Seliverstov was killed, as was Sergeant First Class Antonov. Sergeant First Class Petkevich saved himself by parachuting over his own territory, and Sergeant First Class Ageshin managed to take his damaged aircraft to the nearby airfield Shum, where he made a belly-landing.[64]

This aerial battle had only barely ceased before "Assi" spotted a single La-5 fifteen hundred meters further down. "Daddy's going to teach that one too," he called out to Stotz in his usual manner, pushed the stick ahead and placed a well-aimed volley of bullets into the plane of the Soviet pilot, who had apparently been taken completely by surprise. Then, Hahn climbed triumphantly back towards Stotz, without noticing that this Lavochkin limped away with a smoking engine. The Soviet pilot, who belonged to the 522nd Fighter Regiment, carried out a belly-landing in his own airfield.[65]

Having tasted blood, Hahn and Stotz took off again after lunch, now together with the other two from the "expert group." Landing at the Krasnogvardeysk airbase 50 minutes later, they were able to report no less than nine downings: three La-5 each by Hahn and Stotz, one La-5, and one Curtiss P-40 by Beisswenger, and one Il-2 by Albin Wolf. For whatever reason, however, Soviet primary sources show that the "expert group" was a little too optimistic this time; the only Soviet loss in that battle was one La-5—once again from the 263rd Fighter Regiment—which, with Second Lieutenant Rostem in control, carried out a belly-landing on the Soviet side of the front.[66]

With seven kills—Nos. 80-86—Hahn set a new personal record on this January 14, 1943. The question is only how many of these La-5s that he actually shot down; it could not have been more than five.

Things were the same on January 23, 1943, when the German fighter pilots cracked down on five La-5s from the 2nd Guards Fighter Regiment, as these were trying to get to a formation of German bombers. Colonel Yemelyan Kondrat, the Soviet regiment commander, suddenly discovered how Captain Afanasiy Sobolevs Lavochkin was attacked by a German fighter. Kondrat yawed to come to the rescue of his comrade, but it was too late—Sobolev's machine was already in flames. The next moment, Kondrat was filled with terror as rows of tracers came flying past his own plane, from behind. "Help me, Sasha!" he cried out over the radio to his wingman, First Lieutenant Aleksandr Mayorov. However, he had already driven off the German. Relieved, Kondrat saw how a large parachute canopy unfolded in the sky—Sobolev had made it by parachute.[67]

Sobolev's machine was the only La-5 that the Russians lost in that area that day. But the Germans claimed to have shot down seven such fighters. It could hardly have been a question of any misidentification, since this was the only radial-engine Soviet fighter at that time. Out of these seven, "Assi" Hahn claimed three.

Meanwhile, the Germans lost two fighters. With one of them, shot down by an La-5, Otto Dürkopp, an ace with eight aerial victories, ended up in Soviet captivity.

In reality, the battle on January 23 had been totally different from the one that had taken place nine days previously, when Hahn and Stotz had come across a group of inexperienced opponents. Colonel Kondrat's formation was a kind of a Soviet equivalent to Hahn's "expert group": the 31-year-old Kondrat himself was even more experienced than Hahn or any of his pilots; he had seen action as a fighter pilot in the Spanish Civil War in 1936-1937, against the Japanese over Chalkin-Gol in 1938, against the Finns in 1939-1940, and, from June 1941, in the war against Germany. He was credited with 16 aerial victories, including five during this attack operation, which aimed at—and succeeded with—breaking the siege of Leningrad. Afanasiy Sobolev, who was shot down in a surprise attack but made it, had by then been credited with 13 aerial victories since 1941. Kondrat's wingman Mayorov had five on his account, and would extend that to 27 before the war was over.

"Assi" Hahn's reported successes during these days is not quit in accordance with what he told Trautloft, which he in turn wrote in his diary, "Hahn says that the aerial battles have not become easier at all, but that they are becoming harder every time. Coming from the English Channel, where he was used to relentless and tough aerial combat, he's now telling us that he has to muster all his flying skills to be able to shoot down enemies which in no way are wanting in comparison with the British."

On January 27 at 1043 hours, Hahn was credited with his 100th aerial victory, against an La-5 at 50 meters altitude in the Schlüsselburg area. The documents that have been preserved in the Russian state military archives in Podolsk show that it was once again Kondrat's 2nd Guards Fighter Regiment that he met. It is quite easy to reconstruct the aerial battle: six La-5s led by First Lieutenant Filipp Kosolapov were attacked by two Fw 190s as they themselves were attacking a formation of German Junkers 88 bombers. Kosolapov and two other pilots reported three downed Ju 88s (the German bomber wing, KG 1, actually only lost one). The German Fw 190 pilot, Master Sergeant Rudolf Rademacher, hit an La-5 with a volley of bullets and saw the plane crash. His wingman confirmed this as Rademachers 30th aerial victory. The Soviet pilot, Sergeant First Class Kalinin, did however manage to raise his damaged machine from the nosedive and make it back to the base, where he landed with some difficulty.

Immediately afterwards, two Me 109s entered—Hahn and Stotz—and attacked First Lieutenant Kosolapov. The clock on Stotz's instrument panel was at 1042 hours when he reported Kosolapov's La-5 being shot down. Kosolapov made it, however, but it was probably his steep nosedive in order to escape that made Stotz believe that he had been shot down. First Lieutenant Ivan Skrypnik and Second Lieutenant Konstantin Fonaryov immediately threw themselves into position behind the two 109s, which aborted with a sharp turn. Things segued into a dogfight where the leading 109—apparently Hahn's—was edging in on Skrypnik. But another Soviet pilot, K. V. Fedorenko, came to the rescue and opened fire on the Messerschmitt, which was seen crashing in a billow of smoke. It was actually "Assi" Hahn hurling his aircraft into a steep nosedive to escape his pursuer—upon which the 109's engine let out thick, black smoke. Hahn dove from 1,200 meters down to 50 meters above ground, where he spotted Kosolapov's La-5. Hahn opened fire and called triumphantly over the radio, "*Abschuss!*"—a downing! It was his 100th aerial victory! But in reality, Kosolapov escaped—in spite of having been reported as shot down twice within two minutes, by both the German aces. Meanwhile, Stotz dove after Hahn—with the same billowing smoke, upon which Skrypnik thought that he had hit and downed this "Messer."

Now, Rademacher and his wingman appeared again in their Fw 190s. The Soviet fighter pilots met them head-on, "but," says the Soviet report, "the Fw 190 pilots were very experienced and avoided this dangerous fight and skillfully flew into a cloud." By then, Hahn and Stotz had climbed to a higher altitude again, and were now coming down on Skrypnik, but he was defended by Second Lieutenant Konstantin Fonaryov, who forced Hahn and Stotz to abort.[68]

As we can see, the Russians reported the downing of three Ju 88s and two Me 109s, and the Germans the downing of three La-5s. The true result was one single downed aircraft — the Junkers 88 from KG 1.

When we reconstruct the battle in this way, it is easier to reconcile Hahn's statement about the very hard aerial battles with their results.

On February 19, 1943 the wing was back in Relbitsy. Suddenly, there was a message from an excited Trautloft in Siverskaya, south of Leningrad: "Now the group has achieved its four thousandth aerial victory! It was Otto Kittel who did it! Come over here to Siverskaya! Let's celebrate it, and at the same time your one hundredth and Stotz's 150th aerial victory! Then you'll all go home on leave!" Max Stotz had sacored his 150th victory the day before Hahn's 100th. With his 39th aerial victory, Sergeant First Class Otto Kittel was the hero of the day. He would later develop into the most successful aviator of the entire wing, with 267 aerial victories, before he was shot down and killed less than three months before the end of the war.

But the celebrations had to be postponed because of the developments at the front. For more than a year, the Germans at Demyansk south of Lake Ilmen had been holding an advanced position, penetrating the Soviet lines as an index finger. Eventually, it became impossible to resist the increasingly severe Soviet attacks, and on February 17, 1943, Operation "Ziethen" was begun, the evacuation of this position. This required air cover, and since Hahn's wing was the only German fighter unit nearby, there could be no leaving Relbitsy for a party before this operation had been completed.

The Germans began their evacuation in the worst possible flying weather, and during the first few days, the retreat was protected against Soviet bomb attacks by a whirling blizzard. But in the evening on February 20, the weather cleared, and early the next morning, the staff at Relbitsy was reached by an alarming report about "unusually heavy Russian activity in the air" over the Demyansk "pocket." "Assi" took off together with Max Stotz and their two wingmen. It was his 560th, and—as it would turn out—last combat mission.

Among other Luftwaffe veterans, it is quite well-known that "Assi" Hahn liked to "honor" the custom of having a glass of *Zielwasser*—schnapps—before a sortie. But this morning, it seems as though he had taken one too many. "I wasn't actually that interested in flying this mission," he recounted after the war. "I was tired, I was looking forward to celebrating the group's fourth thousandth downing, and above all, to go home to my wife on leave. I admit that I'd perhaps had one schnapps too many before taking off that morning, and I would regret that."

At the same time as the six aces took off from Relbitsy, six La-5s took off from the Soviet 169th Fighter Regiment from the airfield at Zaborovye on the other side of the frontlines. One of these fighters were flown by Second Lieutenant Pavel Grazhdaninov, one of the Luftwaffe's most dangerous adversaries in this area.

The twenty-two-year-old Grazhdaninov had been selected for his above-average flying skills to become a test pilot at the state-run aircraft factory No. 21 in Gorky. There, he had been testing the La-5 from early 1942, and when he was transferred by the end of the year to his first frontline air unit, the 169th Fighter Regiment, there was hardly anybody who knew the Lavochkin La-5 inside and out as Grazhdaninov did. This also showed immediately. It always used to take the pilots some time to "get into" the aerial battles, but Grazhdaninov shot down two German aircraft already during his very first aerial battle. After scoring another three victories, he was wounded in a collision with a Junkers 88 on December 29, 1942, but now, he was back again, having recovered from his injuries.

Max Stotz's battle report describes how it was the Germans who discovered their opponents first, "At 0909 hours, over the southern part of the isthmus, we discovered eight La-5s at an altitude of 2,500 meters." Hahn and Stotz were 200 meters higher and dove without hesitation down onto the Russians. "Assi" Hahn recounted the event himself, with certain exaggeration:

"We attacked fifty to sixty Russians. Already on the first approach, I had one of them in front of me. One hundred meters, seventy—at a distance of sixty meters I opened fire. All I saw was a great ball of fire, and then, the Russian was gone. In my headset, I could hear Stotzy declare, 'That'll teach him!'" It was Hahn's 108th aerial victory. The Soviet battle report shows that Lieutenant Colonel Michail Vorobyev was killed as his La-5 was shot down during the first attack. "Assi" Hahn placed himself behind the Grazhdaninov—Second Lieutenant Balandin two-ship formation. He shot down Balandin, who bailed out. Hahn continues his story:

"The second Russian turned unbelievably sharply and came around behind me. His not too friendly greetings slapped into my left wing and came dangerously close to my cockpit. I thought that I needed to shake him off first and then 'teach' him. Meanwhile, Stotzy's opponent went down in flames. It was his 160th."

Three Me 109s—Stotz and his wingman, Master Sergeant Wefers, and Hahn's own wingman, Walter Repple—came to Hahn's rescue. "Watch out, Pavel! Three Messers behind you!" Sergeant First Class Aleksandr Davydov called over the radio to Grazhdaninov. Meanwhile, the Soviet division commander, Captain Aleksandr Chislov and his wingman came careering to save their comrade. Stotz and Wefers med them in a swirling dogfight. Meanwhile, Grazhdaninov did a new sharp turn. One can imagine what it felt like for the pilot with all the G-forces that he was subject to, but he escaped his pursuers and was soon on Hahn's tail again.

"I soon noticed that I was dealing with a true expert," Hahn recounted, "because he was making it damn difficult for me. I'd never met such an opponent before, neither over the Channel nor in the East. The harsh aerial battle was, as happens so often, descending in altitude, and soon, we were near the ground, close above the north Russian primeval forest. It became very hot inside my Messerschmitt. The canopy fogged up, so I had to push back the side windows."

Repple, Hahn's wingman, reported afterwards, "I was going to attack an La-5 that was flying a bit lower, but I saw that an Me 109 had already gone after it. I recognized this Me 109 as Major Hahn's, so I climbed, heading southwards. At an altitude of 2,000 meter, there were two aircraft approaching me. I identified them as La-5s. I turned around and dove against the La-5s, and then, I saw them chasing an Me 109."

This account, while a little hard to understand, is perhaps explained in the Soviet report, which says that "the enemy turned around and tried to flee, but to one of them, it was too late."

We will now. return to "Assi's" account—remember that all this played out within perhaps less than one minute: "During the aerial fight, which played out at the lowest possible altitude, I lost orientation. I no longer knew whether I was above friendly or hostile territory. I set off at full speed westwards while two Russian fighters attacked me from behind and from above. Alone, with only a little ammunition left, and in a hot cockpit, I was not eager to continue the aerial battle, so therefore, I gave up my last altitude to gain some speed and placed my Messerschmitt right above the treetops. A few seconds later, there was a loud bang that shook the entire aircraft. At the same moment, the propeller stopped."

By then, Stotz had lost "Assi" out of sight. He heard a dramatic call over the radio from his wing commander and friend, "Stotz, they've taught Dad a lesson! I need to make an emergency landing!"

"Assi" put his machine down in a belly-landing in a field next to a highway southwest of Demyansk. As he climbed out of the cockpit, a whole platoon of Soviet soldiers came rushing, their weapons ready to fire. The German ace showed that he did not stand a chance to set fire to either any papers or his aircraft, so he raised his hands in the air and prepared for the worst.

During World War I, the aviators on both sides used to treat each other with a sort of "chivalrous" respect. It was not unusual for them to fly past and drop a wreath when an opponent was buried, and often, enemy airmen who had been shot down and had survived were invited as guests to their airfield before being taken to a prisoner of war camp. It is quite well-known that the Germans treated their British opponents that way during the aerial war over the English Channel. "Assi" Hahn did that too on several occasions. Perhaps the most renowned case—portrayed in several books and also on film—is when Adolf Galland invited the British fighter ace Douglas Bader, "Tin-Legs Bader," to his airbase in the Pas de Calais area.

The fact that Germans and Russians often did the same on the Eastern Front is less well-known, perhaps because it was not particularly popular to talk about this during the age of ideological contention during the Cold War, neither in the West nor in the East. But the fact is that the captured Hahn was invited to the Soviet airbase at Zaborovye. Instead of being executed immediately, which was what propaganda had made him fear, he suddenly found himself sitting in a car in front of the gates of a Soviet airfield. A fighter pilot named Stepan Mikoyan— he wrote his memoirs after the war, where he mentioned the encounter with Hahn—met up. Mikoyan spoke excellent German and would serve as an interpreter.[69]

Zaborovye was the base of the Soviet 210th Fighter Wing, which consisted of the 169th Fighter Regiment and the 32nd Guards Fighter Regiment. Both were a kind of elite units, especially the 32nd Guards Fighter Regiment, which had risen to prominence during the Battle of Stalingrad, where it had been credited with 201 kills against 53 own losses.[70] In early 1943, most of the pilots of the 32nd Guards Fighter Regiment actually wore medals of various kinds for their merits. Three of the had been give the highest Soviet decoration, the Hero of the Soviet Union, which can be seen as an equivalent of the Germans' Knight's Cross: Captain Andrey Baklan with 39 aerial victories, Vasily Babkov with more than 20, and Sergey Dolgushin with 15. Among the other pilots at the unit, Aleksandr Kotov had 16 aerial victories by then, Michail Garam 13, and Aleksandr Aniskin, Ivan Kholodov, and Aleksey Kholzunov 10 downings each.

The 32nd Guards Fighter Regiment's first commander, Major Ivan Kleshchov, was with 48 aerial victories the most successful fighter ace on the Allied side when he was killed as the transport aircraft that he was traveling in crashed in bad weather on New Year's Eve of 1942.

Appointed as his successor was no-one less that the Soviet dictator's son Vasiliy Stalin.

The latter was, judging by all appearances, done to appease the capricious ruler in the Kremlin. In any case, it was not a very successful decision. Having not yet turned 22, Vasiliy Iosifovich Stalin lived up in all respects to the cliché of the spoiled rich man's son. Now, perhaps he was not quite as spoiled; when his mother passed away when he himself was only eleven years old, his father had handed him and his sister Svetlana over to be raised by a nanny. Vasiliy's upbringing was characterized by tumultuous revolting against this treatment, and he soon learned that he, being the son of the great dictator, could take quite large liberties.

Vasiliy learned to fly, but flew his planes as if they were sports machines, and crashed many of them due to nothing else than recklessness. He never became a very good pilot, since he never listened to criticism, but demanded to become a fighter pilot to win success and fame. In order to protect him, he was placed in the unit with the largest number of fighter aces, the 32nd Guards Fighter Regiment. He had not earned his rank of major any more than his "wings" from the flying academy. When he was informed, on New Year's Day 1943, in a state of a hangover from the previous night's wild partying, that he had been appointed wing commander, he appeared to have felt that he was now almost almighty.

The group commander, Colonel Valentin Ukhov, was pulling his hair out. The 35-year-old Ukhov was a very experienced fighter pilot and a scrupulous unit commander. He had trained as a pilot already in the late 1920s, and in 1935, he logged no less than 1,500 sorties. He participated as a volunteer in the Spanish Civil War, where he shot down four of the Franco side's aircraft, and served after that as head of an academy of advanced flying. In the summer of 1942, he was appointed commander of the newly-formed 210th Fighter Wing.

Ukhov was a very popular unit commander. It was said of him that he knew even the least mechanic at his unit, and that he always had a cheerful remark at hand. At the same time, he was not afraid of expressing his opinions to his superiors—something that would cost him very dearly. Vasiliy Stalin almost immediately came at loggerheads with Ukhov, who had no time for trespasses of discipline—whoever the perpetrator may be. The "brat" Vasiliy Stalin's arrogant manners, undisciplined behavior, and wild parties with women and alcohol were also things that drove Ukhov insane. The conflict between them was somewhat of the talk of the entire Soviet airborne army in the Demyansk sector.

Vasiliy Stalin's appointment as unit commander had a decidedly negative impact on the 32nd Guards Fighter Regiment. The fact that the pilots were very successful in aerial battle did not mean that they were automatically the most well-judged people on the ground, between their missions. Pretty soon, under the young Stalin's command, the 32nd Guards Fighter Regiment had turned into a rowdy party gang with large quantities of alcohol and ladies of pleasure constantly present in the barracks. Ukhov pursued an utterly Sisyphean struggle against this.

It was this quite intoxicated gang that "Assi" Hahn was now facing. He recounted himself, "The door to what seemed to be the officers' mess was opened, and I was brought inside by my interpreter. About ten to fifteen young flight officers, impeccably uniformed and with several decorations on their chests, were standing smoking and drinking at a bar. A young lieutenant with the Gold Star of the Hero of the Soviet Union came and stood in front of me, legs wide apart, and reached out his hand. But I wasn't going to lift a finger for him."

When the other Soviet pilots saw how Hahn treated him, they were upset. The interpreter nervously explained that the German prisoner might want to show him a little more respect, since he was a Hero of the Soviet Union, with seventeen aerial victories. But Hahn was still convinced that he would soon be executed, so he replied scornfully:

"If this hero has been shooting down seventeen fascists, as you claim, he has about one tenth of my victories!" Then, he added, "The Russian pilot who can defeat me has not yet been born. I took out three of you today!"

This encounter has been described both by "Assi" Hahn and a Soviet report. But in the Soviet report, nothing is mentioned about what happened immediately afterwards. According to Hahn, the reaction to his remark was a sheer volcanic eruption. The obviously intoxicated Soviet airmen were absolutely furious and flew at him. They held hm down while one of them climbed up on a chair and strung a telephone wire to a light hook in the ceiling. Hahn paled as he realized that they were going to hang him on the spot. The next moment, they had lifted him up onto the chair and threaded the telephone wire, which they had made into a noose, around his neck.

Suddenly, the door was thrown open. Hahn recounts the sight that met him, "In strode an about fifty-year-old man in a black leather jacket and a uniform hat. He roared something, went straight up to my interpreter, and struck him in the face, and then, he floored that Hero of the Soviet Union with a single blow. The other Russians backed off in terror, while the officer, who turned out to be a colonel, climbed up onto my chair and cut the telephone wire that had been threaded around my neck with a knife." At the very last moment, Valentin Petrovich Ukhov had appeared to save the German prisoner.

Under Ukhov's supervision, Hahn received much better treatment. He was escorted to the mess of the 169th Fighter Regiment, where he was introduced to the regiment commander Nikolay Ivanov. He and his pilots were of a completely different kind than Vasiliy Stalin's feral rabble. They had dinner with Hahn, and then, "Assi" was allowed to sit in one of their fighters. Hahn studied the instruments and controls with interest, wondering to himself whether he would be able to start it by himself and fly away. Ivanov seemed to have read his thoughts. He laughed and let the interpreter translate that starting it was out of the question—the aircraft he was trying out of course had empty fuel tanks.

Major Ivanov was a very experienced pilot who had graduated from the aviation academy at Borisoglebsk as early as in 1931. During the three days that Hahn stayed at Zaborovye, a kind of friendship developed between the two. When Hahn wrote a book a few years after the war about his time in Soviet captivity—*Ich spreche die Wahrheit*, quite a grim book about great difficulties in prisons and prisoners' camps—he mentions that he got along well with Ivanov. Hahn was even able to find an outlet for his interest in sports at the Soviet airbase, where he trained boxing with the Russian airmen.

Over at Relbitsy, the loss of the popular wing commander "Assi" Hahn came as a real shock. "As always, they had tied his large dog 'Lux' to a tree as he took off," Trautloft wrote, "but this time, the dog waited in vain for his master to return."

Several of the men at II./JG 54 had a hard time accepting that their "Major Assi" was gone. Trautloft wrote, "'Perhaps he'll land by us one of these days with a Russian aircraft', they told each other, and so the legends about him already started to spread. This was of course due to Major Hahn himself. There simply wasn't anything that you couldn't believe about him. After only a few days, he had turned into an almost mythical figure among the soldiers."[71]

After a few weeks, they learned more about Hahn's fate: a Soviet pilot from Zaborovye, who had been shot down and captured, told them that the German ace, Major "Gan," with the Knight's Cross and Oak Leaves and "more than one hundred and fifty victories" had ended up in captivity unhurt. The Russian pilot recounted that he had seen him sit down for dinner with the pilots at Zaborovye. After that, he had been transferred to the Vypolsovo airbase, where he had had his own room. According to the Soviet pilot, "Major Gan" was in good health.

In late February, 1943, Hahn was transported to the prisoner of war camp Borovichi, 120 kilometers east of Lake Ilmen. Following a failed attempt to escape in April, 1943, he was locked up in the infamous Lyublyanka prison on Moscow, where he was held for more than a year. In June, 1944, he was sent to the secure unit Block 4 in the Yelabuga prison camp in the Soviet Republic of Tatarstan. Only after the end of the war did Hahn get to a more "humane" prisoners' camp, in Gryazovets northeast of Moscow, where he met other Luftwaffe aces such as Erich Hartmann and Hermann Graf. In December, 1949, Hahn became one of the last German prisoners of war to be freed.

When he came home to meet his dear wife again, seven years later than he had intended, he learned that she had left him for another man. After having subsisted at various jobs that his old comrades helped him to get, he became the head of the WANO Schwarzpulver GmbH gunpowder factory in Kunigund in 1971, and settled in the idyllic Goslar in Harz. The same year, he married Gisela von Vietinghoff, the daughter of the famous general Heinrich von Vietinghoff. Mrs Gisela Hahn was an absolutely charming hostess as I visited them in their home a few years later.

Sadly enough, "Assi" did not have a particularly long life—even if it could certainly have become much shorter. In the early 1980s, cancer got hold of him, and around Christmas, 1982, I received a sad card saying that he had passed away on December 18.

However, the stories about "Major Assi" live on among the shrinking number of Luftwaffe veterans. In 2002, the aviation historian Jerry Crandall published the biography *Major Hans "Assi" Hahn—The Man and His Machines.*

Gisela Hahn remarried in 1983—with the veteran night fighter pilot Wolfgang Falck. He and "Assi" had become friends as early as in 1937, and they and their (then) wives became good friends. Falck's wife had passed on in 1982, the same year as "Assi." But, concerning "Assi" Hahn, the new couple Wolfgang and Gisela agreed, "Nobody says anything bad about him." [72]

Valentin Ukhov, the Soviet colonel who saved "Assi" from an ignominious death on that February day in 1943, fared less well. After Vasily Stalin had accidentally killed a soldier at the regiment while fishing with hand grenades, Ukhov tried to have him kicked out of the air force. A few years later, Ukhov was suddenly put on trial, accused of "anti-Soviet activities." The charges were so absurd that he had to be acquitted in the first trial, but soon, he was detained again. This time, the count was "corruption." Pleading not guilty, he was sentenced to ten years' hard labor. He was freed only after Josef Stalin's death in 1953, and the following year, he was exonerated where the indictments against him were declared false. But his time in prison camps had taken its toll on the, by then, 45-year-old airman's health, and only a few months later, in 1954, he passed away. In an interview soon before his death, the broken man bitterly recounted that "V. I. Stalin has cost me much time and energy."

"Assi" Hahn and his dog "Lux" in France in 1941. (Photo: Hahn.)

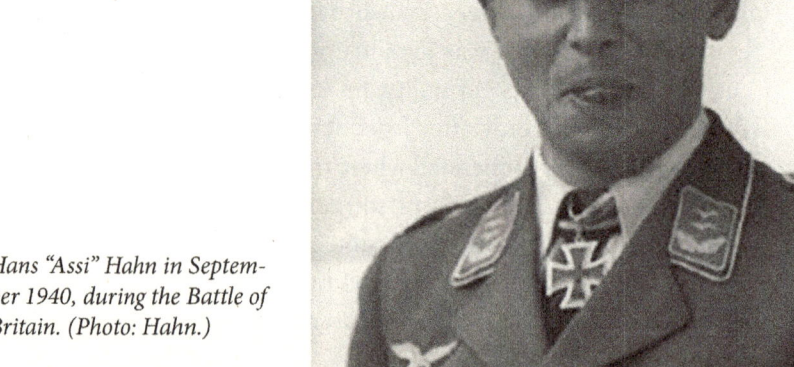

Hans "Assi" Hahn in September 1940, during the Battle of Britain. (Photo: Hahn.)

"Assi" Hahn two years later, on the Eastern Front. The air battles with the Sobiets were not as easy as Hahn had imagined. (Photo: Hahn.)

Max Stotz (front center) and some of his friends, surrounded by admirers in Germany. (Photo: Trautloft.)

Soviet fighter pilot Stepan Mikoyan was "Assi" Hahn's interpreter and guide when he was taken to the Soviet air base Zaborovye. He was the son of Soviet Prime Minister Anastas Mikoyan and later became a famous test pilot. (Photo: Mikoyan.)

Colonel Valentin Ukhov, Vasiliy Stalin's commander, saved "Assi" Hahn's life. (Photo: Mikoyan.)

Joseph Stalin's son, Vasiliy (1921-1962). (Photo: Mikoyan.)

From left: "Assi" Hahn, Max Stotz, and Hans Beisswenger are congratulated on Hahn's 100th and Stotz's 150th air victory in January 1943. On the wall above hang pictures of the Luftwaffe commander Göring (left) and the fighter general Adolf Galland (right). (Photo: Hahn.)

CHAPTER 3

'Daddy' Mölders—the Saint of the Luftwaffe

The circle of Luftwaffe veterans is like all other groups of people—there is talk behind each other's backs, there is plenty of gossip, and opinions about one thing after another. One person is this way, another one is that way, so-and-so is "*ein Arschloch*" (an arsehole), so-and-so is "*ein Pfundskerl*" (a great guy), and so on and so forth. But there is one person who seems to be elevated above all forms of gossip, all forms of negative opinions, and that is Werner Mölders. The question is whether he actually was not—and still is—the most popular aviator in the Luftwaffe.

Werner Mölders really stands out in many ways. He was the German top ace in the Condor Legion, the German air force that fought on Franco's side in the Spanish Civil War. He was the first leading fighter ace of World War II—on both sides; he was the first one to reach 20 aerial victories, 40 aerial victories, and 100 aerial victories. His success as a pilot was so great that Hitler even invented a new highest decoration for valor because of Mölders. In fact, Werner Mölders revolutionized the entire fighter aviation tactics in a way that all air forces still bear the marks today.

But above all, he was somewhat of a father figure for the young German fighter pilots. He took care of them, showed them great solicitude, and taught them both how to survive and become successful. They lovingly called him "Vati," "Dad." When Werner Mölders was killed in an air crash in late 1941, only 28 years old, there was probably no German airman who was not deeply shaken. The question is also whether Mölders did not also turn against the Nazi regime towards the end. We shall return to this.

This chapter is largely based on interviews with and material that has been made available by Werner Mölders' brother Victor, carefully examined against wartime documents.

Werner Mölders was born in an upper middle-class home in Rotthausen in Gelsenkirchen. His father was head teacher at the Schalke high school in town, and his mother a typical submissive middle-class wife with fine manners. But Werner barely had any memories of his father; he was killed in the trenches of World War I on March 2, 1915, sixteen days before the young boy's second birthday.

For the rest of his life, Werner seems to have been seeking a father figure. The first real father figure in his life was the Catholic priest Erich Klawitter, eleven years his senior. He led Mölders to a deep faith in God and membership in the Jesuit-influenced Bund Neudeutschland (BD), which Klawitter himself was one of the founders of. This students' association for Catholic boys (girls, who had initially joined, were expelled) was strictly conservative. The organization's aim was "a new, better, Christian Germany" and it was strongly influenced by the medieval knights' orders as perceived by Romanticism. Victor Mölders recounted what a severe blow it was for the young Werner when Klawitter was called to Berlin for a new ecclesiastical position in 1930. But he continued following the BD's writings faithfully. Hitler's assumption of power in 1933 was hailed by the BD, which declared that it was a positive thing that "the liberal Weimar state had been defeated."[73]

By then, Werner Mölders had joined the military, and that would take him to a considerably darker father figure. 18 years old, he applied in 1931 for officer's training with the *Reichswehr*, where he ended up in Infantry Regiment 2 for basic training. One could probably say that Mölders was subject to bullying at the barracks. The other recruits were teasing him for what they thought was his exaggerated religiousness, and he had to endure being called a "Catholic spy." The grave and unexcitable Mölders neither smoked nor drank, nor did he participate in any parties or beer nights with his comrades, which worsened his isolation in the circle of comrades. Mölders turned out to be quite sensitive and complained to his mother and his siblings about the slavish discipline. It was during his time as a recruit that he arrived at a notion that would later characterize his own style of leadership: "Order needs to be based on discipline based on understanding, a voluntary discipline," he told his younger brother, Victor. He in turn recounted that the time at Infantry Regiment 2 (IR 2) changed Werner very much, it made him much humbler and not as strict and demanding as he had used to be.

After a year and a half, his basic training at IR 2 was over, and Werner Mölders came to the officer's training at the infantry academy in Dresden. There, he was able to breathe freely. Here, he felt much more accepted, here, the atmosphere was much more tolerant.

Out of 142 fellow students of Mölders' at the infantry academy in Dresden, 69 would be killed in World War II. But one of them also fell afoul of the Aryan Paragraph, which forbade "non-Aryans"—that is, Jews—from serving in the armed forces. Another of his fellow students was the son of a nobleman, Günther Freiherr von Maltzahn.[74] The friendship that would develop between Mölders and von Maltzahn would influence Mölders even further to gain a more balanced view of things rather than the simplified worldview that he had been fed in his youth.

But when the Nazis rose to power in 1933, he was still strongly influenced by the Catholic students' association's opinions. "At last, we can breathe freely after years of red government," he wrote in a letter to Klawitter, continuing, "Baptized a Catholic, Hitler is pursuing a struggle on a national and Christian basis. I admire him as I do no other living person today."

However, only a few months later, Mölders' illusions about Nazism would become severely tarnished—when the Nazi state had all non-Nazi youth organizations, including the BD, dissolved. This led to fierce disagreements between him and a few Nazis at the infantry academy. In September, 1933, Mölders wrote in his diary, "First comes my God, then my fatherland, and National Socialism. If Hitler tries to crush the

Catholic religion by force, I will be prepared to lay down my life in battle against him for my Catholic faith."

Mölders dreamed about becoming an airman. So far, Germany was forbidden by the Treaty of Versailles after World War I to own any air force, but led by Hermann Göring, it was busy building one in secret. In November, 1933, Mölders took a test of suitability for air service—which showed that he had a strong tendency towards airsickness! But he still managed to be admitted to the "civil aviation academy" in Cottbus, which was a cover for training military airmen. There, he would soon also conquer his airsickness for good.

In 1935, Hitler's second in command, Hermann Göring, announced the foundation of the new German air force, the Luftwaffe. Werner Mölders was placed in the new Air Wing Schwerin, with the honorary name "Immelmann" after the famous World War ace Max Immelmann. This would develop into a Stuka unit (dive-bomber). In April, 1936, Mölders was promoted to First Lieutenant.

But when the Stuka unit no longer needed any Oberleutnant, Mölders was transferred to fighter group JG 134. There, his above average flying skills were soon noticed, and he was made a fighter instruction already in 1936. From 1938, he was training young pilots on the revolutionizing new fighter Messerschmitt 109.

Meanwhile, the Spanish Civil War was raging. There, the right wing led by General Francisco Franco, with strong support from the Catholic church, had rebelled. Hitler and Mussolini actively supported Franco, and Germany contributed with an air force called the Condor Legion. To the devout Catholic and burning aviator Mölders, there was no doubt about it—when it became possible to volunteer for active duty in the Condor Legion, he did so. In the early summer of 1938, he came to Spain, where he took over the fighter squadron 3./J 88 from its departing commander Adolf Galland. He had been flying old Heinkel 51 biplanes, which had to be used mainly as fighter-bombers since they did not stand much chance in aerial combat against more modern aircraft. But with the Messerschmitt trainer Mölders, a full set of Messerschmitt 109s also arrived at 3./J 88.

Me 109 turned out to be the best fighter of all during the Spanish World War. On July 15, 1938, Mölders scored his first aerial victory. Six Me 109s attacked 25-30 of the enemy's Soviet-made Polikarpov I-15 biplanes. The inferiority in numbers was more than compensated by the fact that the Me 109 was 120 km/h faster than the biplane. Therefore, the German pilots were able to carry out their attacks at full speed, in order to then withdraw as quickly. "I was struggling and struggling sweating like a bull while diving in and out among the agile biplanes and trying to get any of them in my sight," Mölders recounted afterwards. Only after many failed attempts did he manage to get an I-15 in his gunsight and shoot it to pieces.

Instead of celebrating his aerial victory, as other pilots used to do afterwards, Mölders sat down analyzing the battle carefully: Which mistakes had he made? What had he done right? What should he have done? He wrote it all down, crammed it carefully, and took his new experiences with him to the next air combat. That time, things went better. After that, Mölders did the same after each aerial battle: he analyzed every aerial battle immediately afterwards, wrote down his experiences, and discussed them with the other pilots.

But when Mölders shot down his fifth enemy aircraft on August 23, 1938, a twin-engine Russian-made SB bomber, he also came close to being shot down by the bombers' fighter escort. Back on the ground, he drew the conclusion that it was only his own ability to react and his flying skills that had saved his life. But if he had had a chance to call for help from his comrades, it would of course have been much better. His conclusion was that there was a need for radios in the aircraft for two-way communication between the pilots.

The "establishment" within the air force had already agreed that larger aircraft—bombers, transport, and reconnaissance aircraft—had good use for radio. The general notion, not only within the Luftwaffe, was however that the radio equipment, which in those days was quite large and heavy, would have a negative effect on the smaller fighters' maneuverability, but Mölders knew the new high-speed monoplane fighter Me 109 as nobody else did. He argued well for his position:

dogfights are suitable for biplanes that are far too slow to escape an enemy. But with the new generation of high-speed monoplane fighters, this era is over. It had turned out that the Me 109's biggest advantage was its high speed, with which it could quickly dive down onto the enemy, shoot him down, and then set off as quickly again. That was how Mölders had achieved his first aerial victories. Therefore, you could also fit the aircraft with radio equipment.

With the new tactics, based on speed and surprise attacks, and in collaboration between the airmen over the radio, Mölders shot down nine of the opposing side's very agile Polikarpov I-16 fighters with his Me 109 in two months. On November 3, 1938, he scored his 14th and last aerial victory in Spain. After that, he was recalled to Germany.

As the leading fighter ace of the Condor Legion, he was now introduced to two important potentates. One was Pope Pius XI, who was grateful for what this Christian fighter had accomplished for the continued influence of the Catholic church. The other was Hermann Göring.

The air force commander Hermann Göring quickly became a new father figure to Mölders. Göring himself considered Mölders somewhat as the son he never had.

At home in Brandenburg, Mölders was horrified by the traces of the *Kristallnacht* on November 9-10, 1938, the "November pogrom" against Germany's Jews. Amongst other things, his mother's friend, Dr. Lilli Friesicke, had died during the *Kristallnacht,* and Brandenburg's synagogue had been burned down. But Göring assured Mölders that only uncontrolled mobs had done this.

Göring set Mölders to develop new fighter methods.

Quite contrary to what is often being claimed, Göring was very receptive to new thinking, and he gave the young Mölders more or less carte blanche, and it was now that he revolutionized fighter tactics by replacing the antiquated three-plane V formation with a two-ship section (*Rotte*) and a four-plane flight (*Schwarm*), as described in Chapter 1. Göring immediately took this to heart and had it implemented throughout the German fighter command. This became the basis for the German fighter command's unrivalled success in the beginning of World War II.

Already by then, fighter pilots called Mölders their "Dad"—ironically for a man who spent his entire life seeking compensation for his lost father. He was appointed squadron commander, and at his own unit, he attended to the men with great care. It was not without reason that he became a very popular unit commander. But the most important thing of all to them was, of course, that he taught them how to survive and succeed in aerial combat. What he impressed on them was something they had not heard during their fighter training:

"The fastest aircraft is the superior one."

"With superior speed, you can choose superior altitude position…

…choose when to attack

… and choose when to abort."

"Speed—altitude—quick action!"

When World War II broke out in September, 1939, Mölders was commander of the 1st Squadron of fighter wing JG 53. As everyone knows, it started with Hitler attacking Poland on September 1, and two days later, the United Kingdom and France declared war on Germany to defend their Polish ally. But that was not the general opinion in Germany. Influenced as they were by nationalist (not only Nazi) propaganda since the time after World War I, the Germans perceived the war on Poland as a defensive war to regain German territory that the Poles had seized. The demands on Germany in the Treaty of Versailles were thought of as oppressive, and when the states that were guilty of the very harsh conditions of the treaty declared war on Germany on September 3, it led to deep indignation.

Mölders and many Germans saw it as though England out of envy and lust for power wanted to subdue Germany. "These Englishmen are a repulsive people who always drag others into war!" he wrote in his diary. One of the German fighter units introduced a caricature of the British prime minister Chamberlain and the text "*Gott strafe England*" (May God punish England) as the symbol of their unit. (This was a slogan during World War I, ironically enough coined by the German Jewish poet Ernst Lissauer.)

The British and the French, however, did very little to honor their pact with Poland. Along the Franco-German border, soldiers were standing more or less at order arms while Hitler made short work of Poland. It was just as calm in the air, but on September 20, 1939, Mölders achieved his first aerial victory of World War II. It was also a triumph for his revolutionary fighter tactics. He led a smaller group of Me 109s in an attack from superior altitude against six Curtiss Hawk-36 (H-75) fighters from French fighter group GC II/5 "Lafayette." At an altitude of 3,000 meters, where the battle played out, both types of aircraft were roughly equally fast—520-525 km/h—but with the fast attack from above, the Germans were able to shoot down their opponents without any losses of their own. Both French pilots survived. *Sergent-chef* Roger Quéguiner made it by parachute, his comrade *Sergent* Robert Pechaud made a belly-landing.[75] One of them was shot down by Mölders.

With the exception of a few sporadic, minor aerial skirmishes, the so-called "Phoney War" on the Western Front elapsed rather calmly between September 1939 and May 1940. During this period, Mölders trained many future fighter aces, such as Adolf Galland, and led many of them to their first kills. Galland, whom Mölders had succeeded as squadron commander in Spain, described Mölders as quiet, grave, and analytical. It is possible that Mölders displayed this side when in company with the more colorful Galland, but he had become more extrovert with time. Even though he still refrained from smoking and drinking alcohol (at least in any larger quantities), veterans from his unit recount that he was always seen in the circle of comrades at the "casino" (German aviator slang for the officers' mess). On the other hand, he did not visit prostitutes—as opposed to many other of the Luftwaffe's airmen—but was strongly against this.

"I hope that things will soon break out properly against England, because now, we are strong enough to finally sock it to this nation of shopkeepers!" Mölders wrote in the late winter of 1940. On March 2, 1940, he had the chance to personally "sock it to" these Englishmen— even if it was a New Zealander who had to pay the price for Mölders' aversion to the British. A little over a month later, the American news agency AP published the following article:

"LONDON, April 5, 1940. With eighteen pieces of shrapnel in his left foot, and one in his right hand, Flying-Officer E. J. ('Cobber') Kain, the first ace airman of the war to have five German machines to his credit, wears around his neck, attached to his metal Identification disc, a Maori greenstone tiki, or Maori luck charm. He is not superstitious, but wears the tiki for sentimental reasons. He was born in New Zealand twenty-one years ago.

The pieces of shrapnel, which will work themselves through the skin, are mementoes of his last battle over enemy lines. In it he brought down two German machines before a cannon shell, fired by another attacking German crew, set fire to his machine, causing him to jump for his life. It was the first parachute jump he had made. Landing in no-man's land, he dodged towards the French lines, and was escorted behind them by a French officer. Crossfire was blazing as they ran."

On March 2, 1940 "Cober" Kain got in the way of Mölders as he and his comrade Hans von Hahn (not to be confused with "Assi" Hahn) hurled themselves over a formation of Hurricanes from the RAF's 73 Squadron and shot down one each. The pilot of the one that Mölders had shot down bailed out by parachute: it was Edgar "Cobber" Kain.

But that time, Kain made it without any injuries, and was soon in the crossfire again. In fact, he would be shot down by the German ace twice during this month. On March 26, 1940, it was "Cobber" Kain and 73 Squadron that attacked a formation of Me 109s from III./JG 53, which was now led by Mölders. During the first surprise move, Kain shot down two of the German aircraft. But Mölders came to his comrades' rescue, and without difficulty, he shot down Kain, who this time got the shrapnel injuries that the article recounts.

"Cobber" Kain was not any old RAF airman. When he was killed in a flight accident in June, 1940 he was, with 17 aerial victories, the most successful fighter ace on the Allied side. 73 Squadron, which he belonged to, was severely hit by Mölders and his fighter pilots at III./JG 53 during the so-called "Phoney War." In September, 1939, the seventeen pilots at 73 Squadron posed for the photographer; between December, 1939, and March, 1940, nine of them were shot down by pilots from III./JG 53 (see p. 124):

1. Warrant Officer John Winn. Shot down by Hans von Hahn on December 22, 1939.
2. Warrant Officer James "Tubby" Perry. Shot down by Mölders on December 22, 1939.
3. "Cobber" Kain. Shot down by Mölders on March 2, 1940.
4. Pilot Officer "Tommy" Tucker. Shot down by Hermann Neuhoff, III./JG 53, on March 2, 1940.
5. Warrant Officer Donald Sewell. Shot down by Wolf-Dietrich Wilcke, III./JG 53, on March 2, 1940.
6. Flying Officer Newell "Fanny" Orton. Shot down by Arthur Weigelt, III./JG 53, on March 26, 1940.
7. "Cobber" Kain. Shot down by Mölders on March 26, 1940.
8. "Tubby" Perry. Shot down and killed by Ernst Boenigk, III./JG 53 on March 29, 1940.
9. Warrant Officer "Ken" Campbell. Shot down by Mölders on April 23, 1940.

Under Mölders' command, the fighter wing III./JG 53 "Pik As" had considerable success during the "Phoney War" between September 1939 and early May 1940:

In September 1939, the unit had a total of 39 aircraft and 39 pilots. Results:
23 aerial victories distributed across 13 pilots.
Mölders 9 aerial victories.

Losses in combat:
3 own Me 109 shot down.
1 own pilot missing.

On May 10, 1940 the Blitzkrieg broke out in the West. At dawn this day, German troops broke through across the border, pushing forward. The German ground offensive was accompanied by extensive combat in the air above. No-one was more successful than "Daddy" Mölders himself.

On May 15, 1940, he shot down a Hurricane from the RAF's 607 Squadron. With this, the great shipbuilder Smith Docks & Co. in London—with 5,000 employees, and which built 900 ships between 1910 and 1987—lost its owner, the pilot Launcelot Eustace Smith. On May 25, Mölders brought home his 18th aerial victory in World War II, on May 27, he became the first one to reach 20 aerial victories, and two days later, he was awarded the highest award for valor, the Knight's Cross.

Meanwhile, the German ground units had great success. In early June, 1940, the British were forced to evacuate their expeditionary force from Dunkerque. On June 3, the Germans began their final offensive on Paris. During the violent clashes in the air on June 5, when the Luftwaffe aimed at annihilating what remained of the French air force (Operation "Paula"), Mölders achieved two victories, Nos. 24 and 25. He was now the leading ace of the war, on both sides.

Did this make him presumptuous? It is possible that his vigilance faltered. In any case, the French pilot, First Lieutenant René Pomier-Layrargues from the fighter wing GC II/7 managed to sneak up on Mölders with his Dewoitine D.520 and shoot his Messerschmitt 109 E into flames. Mölders bailed out by parachute and ended up in French-controlled territory, where he had a very brutal reception. Had it not been for the intervention by an officer, a group of French soldiers would perhaps have beaten the German aviator to death.

But only a little over two weeks later, France surrendered, and Mölders was able to return to Germany as a free man. After some time of recovery, he was sent to the English Channel with a new mission: he was to lead an entire fighter wing, JG 51. In Winston Churchill's words, the Battle of France was over, and now, the Battle of Britain began.

Excerpt from the book *The Battle of Britain: An Epic Conflict Revisited* by Christer Bergström:
When 48-year-old Oberst Theo Osterkamp took over as Jagdfliegerführer 1 (Jafü 1) on 27 July, his postion as the commander of JG 51 was assumed by 27-year-old Major Werner Mölders. It was Reichsmarschall Göring›s idea that young and 'hungry' fighter pilots as unit commanders would galvanise fighter operations. He decided to start with Mölders.

At 15:00 (German time) on 28 July, Mölders took off from the airfield at St Inglevert to lead his Geschwader on a fighter sweep over southeastern England. At Caffiers, only some ten kilometres away, Major Adolf Galland and the Messerschmitts of his III./JG 26 also took off.

Ten minutes earlier twelve Spitfires from 74 'Tiger' Squadron had been sent up on patrol. They were at 1,800 metres above the rooftops of Dover when the radar stations at Dover, Rye and Pevensey noted that the German formations that had been spotted near Calais had begun to move in the direction of Dover.

Ground control called 74 Squadron and issued directives: Intercept plus sixty Heinkels and forty Messerschmitts spotted by air surveillance just off the coast at Dover![76] The Spitfires climbed in a wide turn to the left. They had reached 5,500 metres when Sergeant Anthony Mould's Spitfire suddenly shook under the impact of 20mm shells. Almost immediately the plane was in flames, and the pilot bailed out. Adolf Galland and his III./JG 26 had been lurking at 6,000 metres, and when the Spitfires came climbing up from below, he had positioned himself in the sharp July sun.

The He 111s had only been a decoy, and now they turned back to their bases. While Galland announced his seventeenth victory (which, however, was not confirmed), his companion Oberleutnant Joachim Müncheberg attacked a formation of Hurricanes from 257 Squadron. Mould's Spitfire had not reached the ground before Sergeant Ronald

Forward's Hurricane was shot down by Müncheberg. In that moment, two other formations of fighters appeared, the Messerschmitts from JG 51 and the Spitfires from 41 Squadron. The latter had been ordered up with eleven machines from Hornchurch at 14:25.[77] The 92's Squadron Leader Hilary Richard Hood was above and in front of the others with three other pilots when he heard a warning over the radio.

Mölders said: 'I flew with my adjutant, Oberleutnant Erich Kircheis. Just north of Dover we met a lower flying group of three Spitfires, and behind that more aircraft appeared out of the haze. We attacked the leading formation.'[78]

A lower flying Spitfire trembled in Mölders' gunsight. The German ace fired his cannons. Parts blown off from the Spitfire came flying towards Mölders, and the British plane began to smoke. But although the Spifire pilot, Flight Lieutenant Anthony Lovell was injured in his hip, he managed to evade further attacks. He steered his damaged machine down towards the base at Manston which was in sight, and after just a few minutes he was able to land and was immediately lifted into an ambulance.

Meanwhile, one of the Spitfire pilots in 74 Squadron had discovered 41 Squadron's exposed position. This was something of the counterpart of Mölders within the RAF, the 30-year-old South African Flight Lieutenant Adolphus Gysbert Malan. Flying Officer John Colin Mungo-Park, second to Malan the main ace of 74 Squadron, said: 'What I like about Sailor is his quiet, firm manner and his cold courage. He is gifted with uncanny eyesight and is a natural fighter pilot. When he calls over the R/T, "Let 'em have it!", there's no messing. The b—s are for it, particularly the one he has in his own reflector sight.'[79] Now Malan attacked in the lead of Red Section. He said: 'Met up with six or nine Me 109s at 18,000 feet coming from the sun towards Dover to attack some Hurricanes. Turned on to the tails without being observed and led Red Section into attack. Gave one enemy aircraft about five two seconds' bursts from 250 yards, closing in to 100 yards. He attempted no evasion tactics except gentle right-hand turns and decreasing speed, by which I concluded he had at least had his controls hit.'[80]

It has long been assumed that Mölders was the pilot of this Bf 109, but it was more likely a machine from I./JG 51. Mölders saw a Spitfire attack a Messerschmitt. He manoeuvred into position behind the British fighter, but this made a sudden and surprisingly tight turn and came up behind Mölders! An examination of existing documents indicates that this Spitfire was piloted by 41 Squadron's Pilot Officer John Terence 'Terry' Webster. Afterwards, he wrote in his combat report: 'I was passed by another enemy aircraft. I fired short bursts closing from 100 to 50 yards. I then saw black smoke coming from the cowling over the windscreen.'

'It rattled violently in my machine', said Mölders. 'The radiator and fuel tank were shot up, and for me there was nothing left but to leave at full speed. I got a whole mass of Spitfires after me.'

In that moment a grey-green Messerschmitt '109 swooped down into position behind Webster's Spitfire. In the '109's tight cockpit sat the ace Oberleutnant Richard Leppla. Just as Leppla took his plane up and pressed the firing buttons, Webster shoved his throttle over the catch. Owing to the Spitfire's ability to temporarily increase the maximum speed by 40-50 km/h he managed to escape his pursuer.

Mölders also escaped and turned back towards his airfield. 'Fortunately the engine held to the French coast', he said. 'Only then did it start to run badly. Then when I was about to land, the landing gear would not deploy. I had to make a belly-landing. As I climbed out of the plane, my legs felt strangely weak, and on closer inspection I discovered large blood stains. The medical examination found three pieces of shrapnel in my thigh, one in the knee and one in my left foot. In the heat of battle I had not noticed any of these wounds.'

Not only did Mölders' wounds render him unable to fly for a few weeks – his own plane was so badly damaged that it had to be scrapped. Even worse, one of the newcomers in JG 51, Gefreiter Martin Gebhart, fell in that battle. Furthermore, Unteroffizier Erwin Fleig had to force-land his damaged '109 on the French Channel coast. The three victories claimed by Mölders and his pilots could not outweigh this. For Werner Mölders it was clear that it would be a tough time at the English Channel. His colleague Major Galland was considerably more positive when he visited Mölders by his bedside; Galland's III./JG 26 had scooped three victories without any losses of their own.

On the British side, 41 and 257 squadrons had had an aircraft each shot down while 74 Squadron had three Spitfires shot down and one damaged. The British presumed that they had destroyed five '109s – all by 74 Squadron."

Mölders would never forget this painful lesson in Spitfire agility.

One of the most famous stories about the Luftwaffe during World War II—popularized through the motion picture *Battle of Britain* from 1969—is how *Reichsmarschall* Göring visits a group of prominent fighter pilots by the English Channel. Questioned about what they wish, Mölders replies that he would like stronger engines in the Messerschmitt planes. Göring promises Mölders that those are on their way, and then turns with the same question to Galland—who, without thinking, replies, "*Geben Sie mir eine Staffel von Spitfire, Herr Reichsmarschall!*"—Give me an outfit of Spitfires, *Herr Reichsmarschall*.

The source of this is Adolf Galland. He wrote about it in his memoirs in 1953—which have subsequently had an almost unmatched influence on the view of the Luftwaffe during World War II—and in numerous statements during the following years. When I asked Galland if that was actually what he said, he replied in the affirmative, but added that he really believed that the Messerschmitt 109 was the better one of the two aircraft; he had just felt so frustrated at the time.

There is another, more independent source for this remark. In my research into reports of interrogations with downed and captured German fighter pilots during World War II in the British National Archives in Kew in southwest London, I came across an interrogation with a German aviator. Questioned what the German airmen thought of the Spitfire, he replied:

"It's a very good aircraft. On one occasion, *Mölders* told *Reichsmarschall* Göring that he wished that his units would be equipped with Spitfires."

During the Battle of Britain, there was fierce competition between Mölders and Adolf Galland over who could shoot down the largest number of aircraft. In mid-August, 1940, Mölders was able to start flying again, after his relatively mild injuries had healed reasonably. Between August 14, 1940, and September 28, 1940, he carried out 51 sorties over the English Channel and achieved 17 victories, Nos. 26-42.[81] During the same period, Galland "put on a spurt" and bagged 27 kills, his personal Nos. 15-41.

This kind of contest was something that was encouraged in the highest degree not only by the air force commander Göring—who was a veteran fighter pilot himself from the "Red Baron's" fighter group during World War I—but also by the German propaganda. It would have been strange if the young, ambitious and adrenaline-filled fighter pilots had not fallen for this.

Most of the time, Mölders was one step ahead of Galland in the contest, even though his rival was catching up. On September 21, 1940, Mölders was awarded the new, highest decoration for valor, the Oak Leaves for the Knight's Cross. General Eduard Dietl, the victor at Narvik received Oak Leaves, and was the first one to be awarded Oak Leaves for the Knight's Cross. Three days later, Galland became the third one. It is obvious that the young fighter pilots were the darlings of the Third Reich.

But the Battle of Britain went badly for the Germans. One of Mölders' groups (a fighter wing normally consisted of three groups; Mölders' wing had four) had to be taken out of combat because of high losses. During the month of September, 1940, Mölders' JG 51 lost one fifth of its entire force. Amongst other things, Mölders lost his adjutant, First Lieutenant Kircheis, and his brother, Victor, who also flew at JG 51. Victor Mölders was shot down on October 7, 1940, and ended up in British captivity.

The German fighter pilot Gerhard Schöpfel, who led III./JG 26 during the Battle of Britain, recounted, "After a few weeks of fighting over the Channel, a new phenomenon established itself within the units. We called it '*Kanalkrankheit*', the Channel Disease. It showed in different ways—either through various psychosomatic infirmities or excessive irritability. In some cases, the infirmity expressed itself as 'technical faults' on the aircraft when it was time to carry out a new sortie over the Channel. Pilots reported faults on their aircraft and aborted the mission. Once back at the base again, the master mechanic was unable to find any fault with the plane, which seemed to have repaired itself in the meantime."[82]

"Mölders was on sick leave and flew no more combat missions. 'The flu' was Mölders' official reason, but that did not stop him from, among other things, meeting with his colleagues among the young Jagdgeschwader commanders in Pihen on 3 November, in Wissant the following day, in Marck on 6 November and in Pihen on the 7th.[83] On the latter day, Mölders, Galland, Lützow, Trautloft and Oberstleutnant Carl Vieck (the former commander JG 3) met for an 'exchange of ideas'. Hannes Trautloft wrote in his diary: 'We are all disappointed with the development of the air battle of the British Isles, but we are particularly annoyed that Göring, who has already said that he could have the fighter aviation disbanded, now puts all the blame for the failure on the fighter pilots. The air campaign lacks thoughtful planning. The incessant changes of operational tactics, from the first massive day attacks by the entire Luftwaffe to the small pinprick attacks by Jabos, which is costing us high losses, has resulted in widespread doubts in our high command out on the airfields. We all say that there have been too many cooks. Although no one says it bluntly, we feel that our optimism, our élan and our fighting spirit has waned. Something has really gone wrong ... Towards midnight we break up, saddened and filled with serious thoughts.'[84]

Of course, the root of the malaise was the defeat the Luftwaffe had suffered at the hands of Fighter Command. The Germans were simply shocked at not winning, and this resulted in a psychological crisis. 'We quarreled with the high command, with the bomber fliers, with the dive-bomber fliers, with the Zerstörer fliers, and finally also among ourselves', wrote Adolf Galland.[85]

Reichsmarschall Göring was also in the midst of a personal crisis, which explains his bad judgement at that time. In a letter to his Swedish brother-in-law Count Eric von Rosen, he spoke of his 'exhaustion'. However, it seems to have been clear to Göring that the fighter-bomber offensive had degenerated into mutual 'face slapping' without any real importance for the further conduct of the war. On 30 October, Göring met with the commanders of the air fleets and air corps in Deauville, where he announced his decision to send the entire fighter force on a 'winter vacation', four Jagdgeschwaders at a time. In practice this meant that the fighter-bomber offensive was cancelled. Göring himself took a 'leave' on 14 November and appointed Generalfeldmarschall Milch acting Luftwaffe Commander. A few days later the Swedish Count von Rosen received a letter from Göring that showed just how bad things were with the Luftwaffe commander's mental condition: 'Right now I am taking a couple of weeks of sick leave because I could not take any more.'" (From *Battle of Britain: An Epic Conflict Revisited* by Christer Bergström, p. 267.)

But the Luftwaffe's staffing policy was quite another story. In spite of the authoritarian Nazi ideology, they knew to use the very latest psychological findings to make the airmen feel as good as possible, to feel comfortable, and to recuperate after a period of difficult ordeals. When Hitler commenced Operation "Barbarossa" on June 22, 1941, Mölders and his JG 51 were back, completely recovered and full of willingness to fight. Mölders, who had been raised into anti-Communism throughout his upbringing, was enthusiastic. Just as exuberant was the Panzer general Heinz Guderian, who led the ground units in the area where JG 51 was operating. "Wherever Mölders is, the air gets cleared!" Guderian said.

In combat with less experienced, technically inferior and tactically neglected Soviet aircraft formations—often bombers without any fighter escort—the German fighter pilots were very successful. Mölders and his wing were now flying the modernized version of the single-engine Messerschmitt fighter, Me 109 F. With this, Mölders shot down five Soviet bombers on June 22, 1941, and another five on June 30, 1941. By then, he—as the second person after Adolf Galland—had been awarded the latest highest award for valor, the Swords to the Knight's Cross with Oak Leaves. But that would soon become insufficient.

When Werner Mölders on July 15, 1941,—three years to the day after his first aerial victory in Spain—achieved his aerial victories Nos. 100-101 during World War II as the first pilot in the world (the number was 115 when including those in Spain), Hitler had a new highest decoration for valor instituted, solely because of this: the Knight's Cross with Oak Leaves, Swords, and Diamonds. Mölders was summoned back to Germany, where he became the first one to receive this fantastic medal.

But now, the German high command felt that Mölders had become far too great for it to risk losing him at the front for propaganda reasons. Therefore, a new position was established—"General of the Fighter Command"—which Mölders was appointed as, a sort of inspector of the fighter command. With that he was banned from flying any front-line missions, but was placed at a staff office in Berlin.

For some time, Mölders has been dating the widow of his comrade Ernst Baldauf, who has been killed in a flight accident, Luise Baldauf. She is five months pregnant when the two are married—by Erich Klawitter—on September 13, 1941.

But it is also by this time that Mölders learns about the true consequences of Nazism. On September 1, 1939, Adolf Hitler had personally issued a written order about commencing the first Nazi Holocaust, the so-called "*Aktion* T 4," which was about mass murder of Germans with psychological or physical impairments. Between 1939 and 1941, 70,000 people were murdered at different hospitals and treatment facilities within the framework of "*Aktion* T 4."

When information about this came out, it caused an outcry within large parts of the German population. One of those who reacted the strongest was the Catholic cardinal Clemens August von Galen, who preached against "*Aktion* T 4" in July and August, 1941. Von Galen was a great influence on the Catholic Mölders, who also learned about other murdering. In a letter, his friend Hugo Satzl told him about the mass murders of Jews, which were begun during the second half of 1941, "Half of the inhabitants in Rozan were Jews. All have gone."

Mölders remembered clearly how his mother's friend had died during the *Kristallnacht,* and he knew from comrades who were of Jewish origin themselves how the noose was tightening around all Jews, including those in Germany All of this was confirmed further before Mölders' eyes in September, 1941, when Germany's Jews were forced to wear the Star of David. Soon afterwards, the first deportations of German Jews eastwards took place.

Now comes the question that has been thrashed over ever since 1941: Did Mölders take a stand against the Nazi regime?

In West Germany, the narrative of Werner Mölders as something of an anti-Nazi resistance man was carefully nurtured. throughout the Cold War. A destroyer and an air unit were given the honorary name "Mölders." Several Luftwaffe veterans tell the same story: how Mölders, after learning about the Nazi murders, wrapped his military awards into a parcel and sent them to the Nazi Party's headquarters in Berlin, with a letter where he declared that he "hereafter refuses to bear the emblems of this regime of criminals," and that he intended to stay with his dear fighter pilots at the front from now on.

The degree of truth in this cannot be verified, and lately—after the fall of the Berlin Wall and the rise of right-wing extremism in Germany—the pendulum has swung the other way. The military history research department in Potsdam established, in an evaluation of Werner Mölders in 2004, the following:

"a) The available material concerning Mölders is so scant that it could be used as a basis to try to prove several different versions.

b) Mölders could be considered the template for a Nazi conforming soldier. His career of service as well as his life story both confirm that he could be taken advantage of by the powers of the day. In addition, Mölders has allowed himself to be used in this way.

c) As can be expected, the person Mölders displays a certain ambiguity. For example, it has been clearly established that he had strong

Christian Catholic bonds. But this does not seem to have led to any considerable distance to the Nazi regime."[86]

What we do know is that Mölders, from September 1941 onwards, exclusively stayed among fighter pilots at various frontline units in the East. When Adolf Galland took over the position in December 1941 as "General of the Fighter Command," he found the staff headquarters in Berlin completely neglected by Mölders. We also know from testimonies by veteran fighter pilots that Mölders flew "illegal" combat missions on the Eastern Front as late as in 1941, thereby winning some "unofficial" air kills.

All this is indeed circumstantial evidence proving nothing. But one thing that the military history research department in Potsdam does not seem to have had access to, is unofficial private photographs of Werner Mölders among his fighter pilots on the Eastern Front late in the fall of 1941. Günther Behling flew, with the rank of lieutenant, an Me 109 with the fighter wing JG 77 on and near the Crimea Peninsula on October-November 1941. Not only does he recall very clearly how Mölders was flying combat missions together with him and the other pilots, but he also took a series of six pictures with his own camera of Mölders at the airbase in the Crimea in November 1941.

In the thousands of photographs of German fighter pilots in uniform that the author has seen during the course of half a century, there has never been any pilot who has not been wearing the decorations for valor that he has been awarded. The German pilots were infinitely proud and even flew sorties wearing the Knight's Cross around their necks, in spite of this causing them some physical discomfort. The Iron Cross, which was attached to the tunic, was always there—for the simple reason that it was compulsory to wear it if you had been awarded it.

Günther Behling's private pictures of Mölders at the Crimea airbase in November 1941 are absolutely unique among all other photographs of German aviators during World War II—Mölders is not wearing a single one of his military awards. In all of the photographs, Mölders' tunic is clearly visible, and in all of these, he has clearly removed both the Iron Cross and all other military awards. The only symbol he is wearing, apart from his rank insignia, is the sewed-on Luftwaffe Eagle, showing that he is an aviator.

What happened next? On November 17, 1941 the Luftwaffe Lieutenant General and Mölders' personal friend Ernst Udet committed suicide. On November 22, Mölders embarks on a twin-engine Heinkel 111 which is to take him to Berlin and Udet's funeral. The aircraft crashes near Breslau. Mölders is dead.

A body was allegedly seen falling from the aircraft before the crash, the aircraft allegedly exploded in mid-air because of a bomb that the Nazi security service supposedly smuggled aboard, Mölders is rumored to have been seen in a Nazi concentration camp, etc., etc. These rumors are being exploited thankfully during the war by the British propaganda, which drops leaflets with forged letters claimed to have been written by Mölders.

The rumor about Mölders in a concentration camp cannot be confirmed in any way, and moreover, several thousands of people get to see his dead body lying in state in the hospital chapel.

But something that fuels various rumors even further is the fact that neither his wife, nor anyone in his family, are allowed to see the dead body.

I asked his brother who was in Canadian captivity when Werner Mölders' aircraft crashed. It turns out that Victor Mölders had opened a small private investigation of his own to get to the bottom of the circumstances around his big brother's death. Victor Mölders tracked down two of those who were aboard the Heinkel 111 and who survived the crash.

One of them is Paul Wenzel, born in 1887. Wenzel was a fighter pilot during World War I and was credited with ten aerial victories. With the rank of major, he becomes Mölders' adjutant during World War II. The other is the radio operator aboard the aircraft, Master Sergeant Arthur Tenz. Both of them are able to tell Victor Mölders the same story:

Werner Mölders took a seat to the right next to the pilot, First Lieutenant Georg Kolbe. Behind them are the flight mechanic, Master Sergeant Gerhard Hobbie. Major Wenzel is sitting next to the radio operator Tenz by the radio further aft in the aircraft.

Somewhere near Breslau, one engine suddenly starts shaking, and finally stops. First Lieutenant Kolbe asks Mölders what he thinks is best—to make an immediate emergency landing or set the course for the airport at Breslau? But Mölders leaves it to the pilot to make the decision. Suddenly, the other engine also stops, and the aircraft is losing altitude. The pilot now needs to perform an emergency landing, but since the machine is steering itself straight towards a tall factory chimney, he is forced to make a sharp evasive maneuver. With that, the aircraft loses lifting capacity at the low speed and crashes to the ground from low altitude. The impact is not too violent; Tenz escapes with a broken ankle, and Major Wenzel breaks a leg and an arm and gets a concussion. But two of the three men at the very forward part of the aircraft are killed immediately. Mölders is still alive, but his spine is broken and his ribcage is shattered. He is taken, seriously wounded, to the Reserve Hospital IV in Breslau, where he dies after a few hours.

Wenzel and Tenz are unanimous in their accounts: It could not have been a matter of sabotage. The engine parts from the aircraft are well-kept enough for the subsequent investigation to conclude that they showed damage due to wear, probably because of lengthy use of the aircraft (this machine was one of the bombers in bomber wing KG 27 on the Eastern Front) in combination with overstraining them during the unusually long flight from Crimea.

Paul Wenzel dies in Düsseldorf on January 30, 1964, six days after hist 77th birthday. Arthur Tenz's further fate remains unknown.

On January 19, 1942, barely two months after her father's death, Werner Mölders' daughter Heidemarie Verena is born. Her mother, Mölders' widow, Luise Mölders, eventually remarries. She has a long and eventful life. Her daughter emigrates to the USA, where she changes her name to Buchanan, and dies the day after her 40th birthday in Alameda, California. When Mrs. Luise Petzolt-Mölders—previously

Baldauf—dies on April 21, 2011, age 98, she is the widow of three husbands—all of them German aviators. Two of them were killed in air crashes, the third one died a natural death.

A film clip showing Werner Mölders as the commander of III./JG 53 "Pik As" in late 1939 or early 1940.

Werner "Daddy" Mölders. (Photo via Victor Mölders.)

Werner Mölders introduced the tactical formations two-plane section (Rotte) and four-plane flight (Schwarm) in the Luftwaffe. (Photo via Mombeek.)

No. 73 Squadron, RAF, in France in September 1939.
Standing far left: Sergeant James "Tubby" Perry, shot down by Mölders on December 22, 1939 and shot down and killed by Ernst Boenigk, III./JG 53 on March 29, 1940.
Standing at the top, fifth from left: Sergeant Donald Sewell, shot down by Wolf-Dietrich Wilcke, III./JG 53, March 2, 1940.
To the right of Sewell: Sergeant John Winn, shot down by Hans von Hahn on December 22, 1939.
Below Sewell, bare-headed: Flying Officer Newell "Fanny" Orton, shot down by Arthur Weigelt, III./JG 53, March 26, 1940.
Top, second from the right: Sergeant "Ken" Campbell, shot down by Mölders on April 23, 1940.
Below Campbell, in a forage cap and with his hands to his mouth: Pilot Officer "Tommy" Tucker, shot down by Hermann Neuhoff, III./JG 53, March 2, 1940.
Standing in the entrance of the shelter, with the pilot's hood: "Cobber" Cain, shot down by Mölders on March 2, 1940 and March 26, 1940.

The Battle of Britain meant a difficult ordeal for the German pilots. This bullet-riddled Messerschmitt 109 E managed to limp back to the French coast fairly and evenly. (Photo: Trautloft.)

Luftwaffe commander Hermann Göring (right) became a father figure for Werner Mölders. (Photo via Victor Mölders.)

Werner Mölders, in turn, became a father figure for many young fighter pilots. Here he is seen together with 21-year-old Corporal Franz-Josef Beerenbrock in JG 51 on the Eastern Front 1941. Beerenbrock had a tally of 117 air victories when he was shot down and captured on the Eastern Front on November 9, 1942. He returned from captivity in 1949 and passed away in 2004, at the age of 84. (Photo: Broschwitz / Traditionsgeschwader JG 52.)

The German fighter pilots reaped great successes on the Eastern Front in the summer of 1941, especially against Soviet bomber planes without any fighter escort. The image shows a destroyed Soviet Tupolev SB bomber. (Photo via Mombeek.)

Werner Mölders in the cockpit of his Me 109 F-4 shortly after he has landed after scoring his 100th air victory in World War II. The Iceland Falcon was the symbol for JG 51. (Photo: Broschwitz / Traditionsgemeinschaft JG 52.)

As a fighter general, Mölders studies the map during an inspection on the Eastern Front. Sitting next to Mölders is Günther Lützow, who became the second pilot to achieve 100 air victories. (Photo via Victor Mölders.)

Werner Mölders and his future wife, Luise Petzold. (Photo via Victor Mölders.)

Günther Behling's private photos of Werner Mölders at the Crimean air base in November 1941 are completely unique among all other photographs of German pilots during World War II: Mölders does not carry any of his militart awards. All photographs clearly show Mölders' uniform tunic, and on all of these he has clearly removed both the Iron Cross and all other military awards. The only symbol he carries, besides the rank patches, is the attached Luftwaffe eagle, which shows that he is an aviator. (Photo: Behling.)

Erich Klawitter marries Werner Mölders and Luise Baldauf on September 13, 1941. At this time, Luise is pregnant in the fifth month and gives birth to the couple's joint daughter after her husband's death. (Photo via Victor Mölders.)

CHAPTER 4

The Terrible 'Jochen' Marseille

One German airman who could perhaps be compared favorably with Mölders concerning popularity among the veterans is "Jochen" Marseille. If Mölders was the "first" among the fighter pilots, there seems to be a kind of consensus that Marseille was the "biggest." Judging by what is being told about Marseille, there could hardly have been many who were more gifted as fighter pilots than Hans-Joachim Marseille. There are, as we shall see, several similarities between Marseille and Mölders—but when it comes to personality, the difference between the two of them could hardly have been bigger.

Adolf Galland has called Marseille the "ace of the aces"; he described him as "the best pilot in the Luftwaffe … and the worst officer!"[87] When Marseille was killed in a plane crash—just like Mölders—only 22 years old, he had been credited with 158 victories, most of them against British fighters, including 17 in one single day. He had also driven several superiors hopping mad.

Born on December 13, 1919, in Berlin, "Jochen" grew up to become a rebel. His father, Siegfried Marseille, a former military officer who had changed careers and become a policeman, was authoritarian and demanding in the most old-fashioned way. This did not suit the young Hans-Joachim, who was a sensitive child, much more like his mother. The difference between Siegfried Marseille and his wife Charlotte could not have been greater, and after only a few years' marriage, they went their separate ways. Hans-Joachim connected very closely with his much more cordial mother, and continued rebelling both against his father and other authorities. Among his teachers in school, he was already known as "The Terrible."

Hans-Joachim was, in all respects, a freethinker. There was no authority or "foursquare" rule that he did not challenge one way or another. He was clearly influenced by the new, free cultural life in Berlin of the 1920s—and then, he was also a dyed-in-the-wool "Berliner" with all that that entails of rebelliousness and humor. Berlin's "underground scene" with jazz music captivated him as a young man. After that, he remained a great music lover. He was a talented piano player and worshiped musicians such as Scott Joplin and Robert Johnson—something that was seen with disapproval by cultural conservative circles, especially the Nazis who came to power as he had just become a teenager.

It is easy to imagine what Hans-Joachim thought of the Nazis' demands for closing ranks. He continued living a free life and associated with several young Jewish boys, in open defiance of the aggressive Hitlerjugend groups. The fact that he was a favorite among the girls was frowned upon by some.

Apart from music, Marseille had a great passion for aviation. Being the romantic that he was, he dreamed about flying freely up there among the clouds. It was his desire to train to become an aviator that eventually made him start taking school seriously. To the great surprise of his teachers, Marseille became one of the youngest, only 17 years old, to graduate from high school.

After having tormented himself through five months the compulsory "labor service," he applied to become and was admitted as a cadet in the Luftwaffe. In the spring of 1939, he began his flight training. But there, it seemed as though he could not hold back any longer. Once again, his "terrible" side reappeared—at least according to his superiors. Time and time again, he was confined to barracks for various breaches of discipline, and on more than one occasion, he went AWOL because there was some local young beauty attracting him. His promotion to Lance Corporal was postponed by him breaking formation on one occasion to carry out a series of breakneck aerobatics. But everyone, including his officers, were forced to admit that Marseille was a divinely gifted aviator.

Klaus Häberlen, who was one of Marseille's flight instructors, recounted how Marseille on one occasion disappeared from a practice flight. A little later, he called the airfield, informing them that he had had an engine fault and had been forced to make an emergency landing "somewhere in the countryside," but that he could fix the fault himself and would return the next morning. When Marseille had landed at the base the next morning, Häberlen had a couple of mechanics examine the aircraft's engine. They found no fault, so Häberlen summoned Marseille to himself. Without further ado, he happily admitted that he had wanted to impress a young woman by landing in a field, and when he was then going to take off from there, it was too dark. Häberlen, who was very impressed by Marseille's flying skills, decided to leave it at that. However, Häberlen was able to conclude that Marseille was "not suitable for blind flying." His assessment of Marseille was:

Bomber pilot—not suitable

Transport pilot—not suitable

Reconnaissance pilot—not suitable

Fighter pilot—"excellent, providing that he gets an understanding commander."[88]

Therefore, Marseille was transferred to a fighter pilot school, Jagdfliegerschule 5, when his basic flying training was completed. There, he had the same assessment: "Marseille is far from any model soldier—he is far too undisciplined for that—but he is a great aviator and will be able to go far, providing that he gets an understanding commander."

In July, 1940, Marseille graduated from the fighter pilot school with top grades. Soon afterwards, he was transferred to the fighter group I. (Jagd)/Lehrgeschwader 2, which was equipped with Me 109s. This Lehrgeschwader—learning wing—was a special unit, originally intended to try out new methods in aerial warfare. His squadron commander at LG 2 was the First Lieutenant Herbert Ihlefeld, five years his senior, who had been flying with Mölders in Spain, where he had achieved seven aerial victories. Ihlefeld was described by other veterans as harsh and not particularly indulgent concerning breaches of discipline. Marseille was not the only one of his subordinates who fell out with him.

During one of his first missions over the British Isles—this was during the Battle of Britain—Marseille broke, without informing his comrades, the formation as soon as he spotted a British fighter and engaged a Spitfire in a four-minute dogfight on his own. He claimed to have shot it down, but was then forced to flee with at least one other Spitfire on his tail. By flying closely above the waves at full throttle, he managed to get back to base, where he made a fine landing.

Ihlefeld was so furious that he needed twenty minutes to calm down before being able to receive Marseille and read him the riot act. Leaving the formation in this way, only to go and "play" by himself, was one of the worst things a fighter pilot could do. "Marseille," Ihlefeld said, "if you ever break formation in that way again, I will personally shoot you down!" Having calmed his emotions somewhat with a glass of cognac, Ihlefeld said that he had been joking about shooting down one of his own, but if it happened again, he would make sure that Marseille was sent to the infantry.

Marseille was indeed perhaps the best pilot at the unit—in August and September, 1940, he shot down six British fighters—but if we look at what he did with his own Messerschmitt 109s during the same period, the image of his success pales:

September 2, 1940: Me 109 E-4 "White 14" WNr 3579 belly-landed, French coast
September 11, 1940: Me 109 E-4 "White 4" WNr 3579 destroyed during crash landing
September 18, 1940: Me 109 E damaged in combat
September 23, 1940: Me 109 E-7 WNr 5094 shot down in the English Channel

In fact, Marseille alone represented more than half or the group's aircraft losses during the Battle of Britain! Moreover, he was undisciplined, and was more interested in adventures with women than obeying orders. In his soldier's book, the following was noted:

On stern rebuke
Three days' arrest
Five days' arrest

One on occasion, Ihlefeld was so upset with Marseille's self-indulgence that he had the young airman summoned, and tore up the piece of paper that would have given him his promotion in his presence. Finally, Ihlefeld had had enough, and asked to be relieved of this good-for-nothing. Marseille was transferred to a squadron in JG 52 under First Lieutenant Johannes Steinhoff—with less successful results.

All veterans who knew them both agree that the combination of Marseille and Steinhoff as commander was the worst possible. Even if his post-war memoirs give another picture, Steinhoff was known to be very picky with formal discipline. Marseille was not only the only German fighter ace who would be "put in the doghouse" by this demanding officer.

It started very badly. Marseille came too late to his new unit, so Steinhoff was already upset when the new aviator arrived. "He came sauntering in as though it were a beerhouse," Steinhoff recounted. Steinhoff roared at Marseille to stand to attention, and then started yelling at him. He waved the soldier's book with all its complaints and demanded an explanation, but only got a grin for a reply. When Marseille then explained why he had not reported to the unit in time, Steinhoff almost lost it. Steinhoff recounted, "He had been with a girl at a hotel and had simply forgotten what time it was!"[89]

On his first mission with Steinhoff, Marseille broke the formation and attacked a Spitfire alone. As fate would have it, he had his own aircraft damaged, and crashed it on the French coast. He was confined to barracks for a week, but leaped out of a window to go AWOL, stole Steinhoff's staff car and went into town, was gone all night, and returned to base in the morning, very intoxicated... and accompanied by two scantily clad women!

With even more notes in red in his soldier's book—including Steinhoff's assessment "unfit for officer duty" –Marseille was kicked out of this unit as well. He was now transferred to the next unit, the fighter group I./JG 27. Here, for the first time, he had a superior, group commander Eduard Neumann, who knew how to deal with the unruly young airman.

Neumann remembered his first encounter with Marseille: "He was dressed in a uniform shirt that was so worn out, wrinkled, and washed out that it was reminiscent of today's jeans jackets; his hair was long, he beamed of light-heartedness, and gave an altogether purely civilian impression, in spite of wearing a uniform."[90] But the wise "Edu" Neumann was the kind of person that Klaus Häberlen had had in mind when he wrote that Marseille could become an excellent fighter pilot, "providing that he gets an understanding commander."

Instead of confronting Marseille, Neumann, eight years his senior, showed understanding and made an effort to bring Marseille to reason in a fatherly way. And there, he struck a chord with the young "Jochen"— he who had been used to an authoritarian father (who advanced during the war to become a general) and unappreciative officers. On the other hand, the commander of the 3rd squadron of JG 27, which Marseille was placed in, First Lieutenant Gerhard Homuth, did not have particularly much patience for the young airman's vices. Moreover, Homuth, as it would turn out, had a tendency towards jealousy of Marseille for his flying skills.

The winds of war swept the fighter group I./JG 27 into North Africa. The Italian dictator Mussolini's quite gratuitous attack from Italian Libya on the British-controlled Egypt in the fall of 1940 had ended in blistering defeat. The Germans were forced to come to their Italian ally's rescue. In February, 1941, Rommel's Africa Corps arrived, and soon afterwards, I./JG 27 was tasked with supplying the Africa Corps with fighter cover against British air raids.

The 3rd squadron of JG 27 made a stopover with their Messerschmitt 109 Es in western Libya. Homuth ordered Marseille to fly to Sirte and ensure that the unit had enough gasoline for the continued flight eastwards. But the hours went by and no Marseille turned up. Eventually, it got too dark, so the pilots had to spend the night on site. Not until the next morning did Marseille appear—although while coming with a lorry with barrels of gasoline, he was without his aircraft; he had done an emergency landing with it because of an engine fault. When the aircraft had been filled up and were ready for takeoff, Homuth turned to Marseille with a forbidding face, saying, "You have crashed your aircraft, now you make your own way to the frontline airbase!"

The distance to the airfield at the small coastal village of Gazala, sixty kilometers west of Tobruk, was 500 kilometers as the crow flies. The Messerschmitts had to stop over several times on their way there, but at three o'clock in the afternoon of April 22, 1941, they finally touched down at their destination.

The climate in the hills by the coast to the northwest was subtropical with green vegetation, but down here, the sand and stony desert extended all the way to the sea. The main quarters consisted of tents. But the beach was right next to the airfield, and from there, a cool wind blew over land.

The pilots had barely begun unpacking their personal belongings before a call was heard, "Marseille is back!" Incredibly enough—only two hours after the aircraft had landed, Marseille came driving a motorcar, and not just any old motorcar. It bore a general's flag! It turned out that he had managed to persuade an army general to borrow his personal car to get to the airfield. Apparently, it was a general who was seriously fed up with constantly being bombed by the British air force, which held air supremacy in North Africa by this time.

But instead of being commended by Homuth for managing to make it to the airfield so quickly, Marseille only had a grunt from his squadron commander.

The air war in North Africa was different in many respects from the one by the English Channel during the Battle of Britain. The most important difference was that the British and their allies dominated the airspace here, because of their superiority in numbers. This had not been as big had a large part of the Italian air force not been bound to the ground by a chronic shortage of fuel, ammunition, and spare parts—the supply lines across the Mediterranean and along the long road from Tripoli in western Libya was and remained the Achilles heel of the Axis Powers in North Africa. Until the fall of 1941, I./JG 27 would remain the only German fighter unit in North Africa; its significance is evident from the fact that more than 20 of the 28 aircraft that the British lost in combat in North Africa between April 20 and May 20, 1941, were shot down by its Messerschmitts.

But the air war in North Africa was—as was the ground war—of quite limited proportions. I./JG 27 with its less than 30 Me 109s were pitted against only about 100 fighters and bombers on the British side.

Another difference from the situation by the English Channel was that here, neither side was aided by radar observations, and neither did the British have any access to any Spitfire planes—which had been reserved for the defense of the homeland—until during 1942. The role that the Spitfire had had during the Battle of Britain was now being played by the American-made Curtiss P-40 Tomahawk, which was equal to the Germans' Messerschmitt 109 E.

The 3rd Squadron of JG 27 carried out its first combat mission in North Africa as early as on April 22, when the task was: escort for Stuka dive-bombers against the port of Tobruk. Above the beleaguered town, British Hurricane fighters appeared. These came from RAF 73 Squadron, led by the fighter ace Squadron Leader Peter Wykeham-Barnes. He, who had already been credited with downing 17 enemy aircraft (six of which he had shared with other pilots), threw himself straight onto the German fighter escort. Sergeant Werner Lange, one of Marseille's comrades, did not stand a chance of escape; his Me 109 crashed, completely enveloped in flames, to the ground. Immediately afterwards, Wykeham-Barnes took on a Junkers 87 Stuka.

But the German fighter pilots soon managed to turn the tables. When the fighting was over, four Hurricanes from 73 Squadron had been shot down, and a fifth one had been seriously injured. Only two of them made it back to base unhurt, and one of these was shot to pieces when standing on the ground by a Messerschmitt 109. The Germans also claimed four kills, one of them by Marseille—which, incidentally, became the 3rd Squadron's first aerial victory in North Africa.

It is difficult to establish who shot down who in this aerial battle, but the ace Wykeham-Barnes had been forced to bail out by parachute from his shot-up Hurricane and landed in the harbor. It is possible that his Hurricane fell victim to Marseille.

The next day, April 23, Marseille was out on yet another mission to escort Stukas to Tobruk, and once again, a formation of Hurricane fighters appeared. Marseille dived towards the Hurricanes, aiming to get in behind one of these. What he did now know was that this one was flown by perhaps the best among all Allied pilots in North Africa by this time, the Frenchman Second Lieutenant James Denis.

The 35-year-old James Denis had been a fighter pilot since 1929. After the surrender of France in 1940, he flew over a four-engine Farman 222 to England, where he was eventually deployed in the RAF. On April 7, 1941, he arrived at 73 Squadron in Tobruk, and now, a meteoric career as a fighter ace began: on April 14, he won his first two aerial victories, followed by the downing of two Ju 87 Stukas—one of them shared with another pilot—and three aerial victories on April 22. When he now saw the Messerschmitt come diving behind him, Denis waited, cool as a cucumber. At the last moment, he sideslipped, so that Marseille's volley of bullets passed him by through the air. Then, the '109 came flashing by. A quick maneuver and Denis had the plane in his gunsight. Marseille felt his Messerschmitt shaking as the bullets hit, and it was only by hurling his machine into a steep dive that he could get out of the line of fire.

But the aircraft was hopelessly lost. Marseille managed to take it down into a belly-landing where the machine was destroyed. The pilot, however, was unharmed.

Marseille's early days in North Africa were reminiscent of those during the Battle of Britain. The following weeks, he shot down three more aircraft... but was shot down himself again on May 21, 1941... once again by James Denis! It seems completely unlikely, but the fact is that Marseille's Messerschmitt 109 was the only aircraft that the Axis Powers lost in North Africa on that day, and Second Lieutenant Denis was the only one on the Allied side who claimed downing any enemy aircraft. Once again, things happened the same way as four weeks earlier. Denis recounted himself:

"As we were approaching the target, I dove steeply but noticed that my wingman was falling behind. [Pilot Officer Jean] Pompei was a very skilled pilot, but he hadn't had any training as a fighter pilot. The fact that he was trailing so far behind was worrying me—not without reason, because when I looked behind me, I saw an Me 109 attacking him. I didn't have a radio and therefore I couldn't warn him. His aircraft was hit by the German's fire, and then, the Me 109 came careering after me. I pretended that I hadn't seen him, but kept my eyes on him all the time, and when he had come within range, I quickly cut the throttle and sideslipped to the left. Since I had had a very high speed, my Hurricane reacted violently. I saw how the volley of bullets passed me by to my right. The Me 109 was unable to slow down as quickly as I did, but came flying past me and ended up right in front of me. Now, a dogfight began, where the Hurricane had the advantage because of its good agility. My aircraft flew straight upwards, hanging by its propeller, as I spotted the Me 109 in the sun. I fired a volley of bullets at such close a range that we almost collided. I saw how my bullets penetrated his fuselage."[91]

With this, James Denis had 8½ aerial victories. In fact, Marseille's Me 109 became his last air kill; soon afterwards, Denis was transferred to Syria and Lebanon, advancing there to become the commander of the air staff in Beirut. But in spite of his personal success in aerial combat, it is possible that Denis' greatest influence on the air war was that he made Marseille mature.

Once again, Marseille had escaped unharmed, but being shot down twice in less than a month became a kind of a memento mori to him. After the first time, he had more or less been shrugging his shoulders as Neumann had warned him to be more careful, to use his brain also for surviving in aerial combat. Second Lieutenant James Denis would become the one to emphasize those words through action.

Moreover, Marseille had a considerably better squadron commander during the summer, in the shape of First Lieutenant Ludwig Franzisket, who was serving instead of Homuth, who had come down with jaundice. Without doubt, it was Neumann and Franzisket who made Marseille calm down. But there was another important factor behind this as well—in the desert, there were no ladies available...

Neumann displayed his confidence in Marseille by promoting him on June 16, 1941, to Leutnant and then let him go to Germany on home leave "to think things through."[92] Marseille returned as a much calmer and more mature person. He was still as "frisky" and happy, but had a much more professional conduct, both when serving on the ground and in the air.

However, there was one person at the unit who still did not want to think anything good about him, the squadron commander Gerhard Homuth. Between this strict "Prussian" and Marseille, a deepening sense of animosity grew. Homuth was very hungry for glory and was also the most successful ace of the unit. He had attained 22 hard-earned aerial victories by May 1941, when he came down with jaundice. When Homuth returned to the front—at about the same time as Marseille did after his home leave—he had been awarded the Knight's Cross and was eager to expand his tally.

However, young Marseille had hardly begun flying again before shooting down new British aircraft. On August 28, 1941, the 3rd Squadron sighted a formation of Hurricanes. In the ensuing battle, Marseille shot down one and damaged another one. That was the only outcome of the fight. Back at the base, Homuth was able to note that Marseille was catching up by raising his number of kills to 14.

Two days later, Homuth was credited with his 23rd aerial victory, a Blenheim. The problem is that while Marseille's downing on August 28 can be verified in Allied reports of losses, no Blenheim was lost on August 30. It could possibly have been a twin-engine Maryland bomber, which, a Messerschmitt 110 pilot also claimed to have shot down, however.

On September 9, 1941, I./JG 27 clashed with seven Hurricane from the British 33 Squadron. Within six minutes, the Germans reported the shooting down of five Hurricanes, two of them by Marseille. The actual British losses were three Hurricanes. From one of them, Squadron Leader Jack Marsden, bailed out. Marsden was so traumatized by this event that he had to be relieved and sent home to England.[93]

In general, this was a very difficult time for the British air force—especially the fighter units—in North Africa. Their own success was limited, and their losses horrific. Between June and December, 1941, 250 Squadron, equipped with Tomahawk fighters, reported the shooting down of 55 hostile aircraft against 29 own losses in combat, that is, more than twice as many as the squadron's assigned strength (12 aircraft). But the comparison with German and Italian reported losses made by the researchers Christopher Shores, Giovanni Massimello, and Russell Guest, show that 250 Squadron in reality cannot have shot down more than between 14 and 17 Axis planes, which means that the unit lost two aircraft of their own for every single success of their own.[94]

Robert "Bobby" Gibbes, who flew, with the rank of First Lieutenant, a Tomahawk in the Australian 3 RAAF Squadron against JG 27 during this period, recounts, "one minute the air is full of twisting, turning, frantic aeroplanes, and the next minute not a single enemy machine can be seen. The enemy has completely disappeared. You then collect the remnants of your Squadron, count them hastily, then the fires burning below. The feeling is a strange one. Some of those fires down below contain the mutilated bodies of your friends."[95]

3 RAAF Squadron—normally consisting of 12 aircraft—lost 21 Tomahawks in only three weeks. It was only the heavy influx of new aircraft and pilots from the fall of 1941 onwards that made it possible for the Allied air force in North Africa to continue to exist at all.

To I./JG 27, things were looking the exact opposite. During August, September, and October, 1941, the unit was credited with 58 aerial victories with only one single aircraft of its own being damaged in aerial combat. One important reason for this state of things was Hans-Joachim Marseille, who inspired and taught the other pilots. He told his comrades how he used to do it:

"I'll carry out one or two dive attacks onto the enemy's formation and get in each time behind an aircraft that I've picked. But the third time, when the enemy's expecting a repeat of this, I'll dive past them, gaining more and more speed as I dive, and the I'll pull up the machine and loop the loop, do half a roll, and shoot down the next enemy while on my back, and then promptly dive away."[96] Marseille was not only a fantastic aviator—he was also an excellent gunner. After the combat on September 9, where he had shot down two Hurricanes, the armorer found that he had not been using more than 22 shells from each automatic cannon—which, at a rate of fire of 520 rounds per minute, was more or less exactly the equivalent of two and a half seconds' fire.

But meanwhile, the number of operational aircraft at I./JG 27 dropped, because of the failure of deliveries of replacements and spare parts to materialize, from 32 in early August to 10 by late October.

It happened in North Africa, too, that pilots fired on parachuting enemy pilots. During one of his first aerial battles in North Africa, on June 18, 1941, the Australian ace Clive Caldwell watched his comrade Donald Munro being shot dead by a Messerschmitt 109 while helplessly dangling in his parachute. "From that point, Caldwell had no qualms about shooting down enemy parachutists," Caldwell's biographer Kristen Alexander comments.[97]

Marseille does not seem to have belonged to those who felt enough hatred to become guilty of such actions. On September 13, he hit a Hurricane with a deadly volley of bullets. The British plane flared up, and the pilot, Flight Lieutenant Pat Byers, bailed out, badly burned. It has been said that Marseille visited Byers in the hospital, and that he flew over his airfield and dropped a message that Byers was being treated by German physicians. A Messerschmitt 109 flying in and dropping such a message over 451 Squadron's base at Qasaba has been confirmed by the British side. Geoffrey Morely-Mower, another pilot at 451 Squadron, saw this with his own eyes. In his memoirs after the war, he reproduced the contents of the message: "We are sorry to report that Lt. Byers was shot down on Sept. 14 by aircraft of this squadron. He was badly burned while escaping the cockpit. He is now Derna Hospital to recover. We wish to express the regrets of the Luftwaffe."[98]

But Byers' life could not be saved; he died on September 20. This was followed by a new visit from a Messerschmitt 109 in the air over the airfield at Qasaba, and a new dropped message, this time with the regretful information that Byers had died from his wounds.[99]

"As a pilot of the same age as Marseille, I can say," Geoffrey Morely-Mower commented, "that this was an exceptional act of gallantry. An enemy had defied a terrible hail of fire to say that he was sorry about my fiend Pat Byers. That was a damned fine thing to do!"

The animosity between Marseille and Homuth was intensified as Marseille caught up with his squadron commander's position as the leading ace of the group. When Marseille shot down Byers, it was his 17th aerial victory. In addition, he had six unconfirmed claims.[100]

The next day, Homuth bagged his 24th and 25th, while Marseille shot down a Hurricane, once again from No. 33 Squadron. This time, the German victory claims—eight without any losses of their own—correspond with actual Allied losses.

It was clear that there was bitter rivalry in the number of downings between the two of them.

On September 24, Marseille became guilty of one of only three established unfounded claims (that is, where not even a misjudgment can explain why the claim cannot be verified against the opponent's reported losses) by him between April 1941 and May 1942, when he returned early in the afternoon claiming to have destroyed a British bomber. The documents on the Allied side show that nine twin-engine Maryland bombers had begun the South African 21 SAAF Squadron's operations this day, but they were not attacked by any enemy aircraft, and all aircraft returned undamaged to base. Nevertheless, Marseille had the downing of a Maryland confirmed as his 19th aerial victory—which means that another pilot (probably his wingman) had confirmed this. Meanwhile, Homuth shot down a Tomahawk—which turned out to be the one flown by Captain Johnny Saunders from the Australian 3 RAAF Squadron; Saunders made it with minor injuries.

Later the same day, both Marseille and Homuth were present as 3./JG 27 dove from above onto a formation of nine Hurricanes from the South African 1 SAAF Squadron. Marseille was onto them first of them all, and shot the Hurricane flown by Second Lieutenant Melville MacRobert in flames. "Mello" MacRobert bailed out by parachute and was rescued by a comrade who landed and picked him up before flying back to base.

Soon after Marseille had been credited in this way with his 20th aerial victory, his wingman, Master Sergeant Karl Kugelbauer, had a Hurricane on his tail, and according to Kugelbauer, Marseille came to his rescue and shot it down. But at the same time—the clock on the instrument panel in Kugelbauer's aircraft was then showing 1647 hours –Homuth reported the downing of a Hurricane.

Over and over again, the Me 109s dove down onto the Hurricane planes, which tried to protect themselves by forming a defensive circle, a so-called "Lufbery," in order to then take advantage of the excess speed to climb sharply for a new attack with the sun behind them.[101] At 1651 hours, Marseille reported the downing of yet another Hurricane.

The South African captain Cornelius "Corrie" Van Vliet was eventually forced to bail out of his badly damaged Hurricane after it had been attacked during a lengthy dogfight and hit multiple times. The most probable thing is that Van Vliet's Hurricane, in the heat of the battle, was recorded as Homuth's 27th and Marseille's 21st and 22nd victories.

After that, Marseille attacked yet another Hurricane. This was hit and dove away towards the British lines in an attempt to escape. Marseille and his wingman followed suit and ignored Homuth's order to rejoin. Kugelbauer later verified that he had seen this Hurricane crash-land. First Lieutenant B. E. Dold crash-landed in the desert and was picked up unhurt after a while by British soldiers.

During the following weeks, I./JG 27 was not particularly noticeable in the air over North Africa—their few remaining airworthy Messerschmitts were insufficient for that. But on September 27, Homuth flew together with group commander Neumann and attacked two Hurricane planes. The two Germans shot down one each—both can be verified against Allied list of losses.

The Germans' solution to the shortage of spare parts that made the air units' strength shrink was moving entire air units from the Eastern Front to North Africa. The IInd Group of JG 27 was responsible during October and November for most the resistance that the Allies met in the air over the North African theater. But on October 12, Marseille was up in the air again and shot down two Tomahawks as his 24th and 25th aerial victories. That aerial battle cost the Allies a loss of four Tomahawks, exactly the number that the Germans reported as shot down.

Soon afterwards, the men from the 3rd Squadron were sent to Germany to be rearmed and train with the new Messerschmitt 109 version, the Me 109 F. This had weaker armament than the Me 109 E—with only one 15 mm automatic gun, fitted to the nose of the aircraft, instead of the two 20 mm guns fitted to the wings—but it was faster and more agile. All this made it an ideal fighter for skilled pilots such as Marseille. It was with the Me 109 F that he would truly show what he was capable of.

When the 3rd Squadron returned to North Africa, it was as if hell had opened its gates. After the long relatively calm summer and fall, the British had begun a major offensive against the Africa Corps, and the air was filled with aircraft. The Allies had amassed 550 planes—against 171 German and 420 Italian. The Italian pilots were now playing a bigger part than before.

By this time, the Italians had three groups (Gruppo) with the new fighter MC 202 Folgore, which was equal to the German Me 109 F. This also had an impact on the aerial fighting. Between November 26 and December 11, 1941, 120 Allied aircraft were shot down, while the Germans reported that they had shot down 69 and the Italians that they had shot down 72 hostile aircraft. The Italian 9th and 17th Gruppos, equipped with Folgores, were credited during this period with 35 aerial victories against only five own losses in combat.[102]

Marseille and the other German fighter pilots now often flew in collaboration with their Italian allies. On December 5, he was present when Me 109s and MC 202s clashed with ten Hurricanes from the South African 1 SAAF Squadron.

First Lieutenant Noel Milne "Sandy" Sandilands had arrived as a new graduate pilot to 1 SAAF Squadron during the summer of 1941. Since then, he had flown 24 combat missions with a total of 30 flying hours, which was well below the experience that the German aces had. When he perished in his burning Hurricane, Marseille was credited with his 26th and Homuth with his 29th aerial victory. It seems as if Marseille first shot and damaged the South African machine after which Homuth delivered the coup de grâce—unaware that it was the same aircraft.

The next day, Marseille seems to have underestimated his success: he claimed to have shot down two Hurricanes and another pilot reported downing a third, but the truth is that five Hurricanes were lost in that engagement.

On December 7, Marseille came up neck-and-neck with Homuth by knocking down Pilot Officer White from 237 Squadron, which was equipped with Hurricanes. When White was killed, Marseille's 29th aerial victory was a fact. Apparently, the temptation to surpass Homuth was much too large for Marseille; his 30th aerial victory on December 8—reported as a "Curtiss P-40"—cannot be supported by Allied reported losses. But this is also one of only three cases during the entire period between April 1941, and the end of May 1942, when Marseille can be convicted of having made unsupported victory claims. The claims made by Marseille during the following days can all be verified against Allied reports.

SQuadron commander Homuth could not stomach that the undisciplined Marseille was more successful than himself, and Marseille could not resist from teasing him—something that seems to have made Homuth loose his self-confidence. While Marseille Increased his tally with half a dozen during the following ten days, Homuth failed to shoot down more than a single one.

But this was a rough period for Marseille personally, not just because of the toughening air fighting. On his 22nd birthday (his last birthday, as it would turn out) on December 13, 1941, his friend Albert Espenlaub was shot down. Espenlaub ended up in British captivity but was later shot while attempting escape.

Marseille's 35th and 36th aerial victories followed on December 17 in a new encounter with the ill-fated South African 1 SAAF Squadron, which lost four Hurricanes—once again as many as the Germans claimed to have shot down. (Once again, "Mello" MacRobert, whom Marseille had shot down on September 24, was shot down—possibly by Marseille.)

But then, "Jochen" could not take any more. He came down with dysentery and was sent to a hospital in Athens. There, he learned that his beloved sister Ingeborg had been murdered. When Marseille had recovered, he was granted compassionate leave and went to his mother in Berlin to seek comfort for the loss of his sister.

When Marseille returned to North Africa in early February 1942, he had changed. The easy-going "Jochen" had now become serious-minded and he was looking gaunt.

Meanwhile, Homuth had surpassed him now had thirty-eight kills. It seems as if Marseille focused on trying to eclipse Homuth at all costs.

The first day in combat after his return, February 8, 1942, he shot down four British fighters. During the first mission of the day, he claimed two, but the British only lost one Hurricane. In the afternoon, he flew in a formation with Homuth, and contributed two to the total of four German claims; the British lost three Curtiss P-40s of the new

Kittyhawk version. The next day, Homuth came up neck-and-neck with Marseille; both of them now had 40 confirmed aerial victories.

But Marseille did not need to exaggerate to outdo Homuth. During an escort mission for Ju 87 Stukas on February 12, he threw himself over a large formation of Hurricane fighters northwest of Tobruk. Within three minutes, he shot down three of the British and after a long pursuit, he shot down a fourth one. The British losses—five Hurricanes—corresponded exactly with the German claims.

The next day, Marseille was again out escorting Stukas, when he saw two Hurricanes attacking a lower-flying formation of '109s. Once again, it was 1 SAAF's pilots that were subject to the German fighter ace. First Lieutenant Franciskus "Frans" Le Roux probably never saw the danger coming before his Hurricane shook under the hits from Marseille's automatic cannon. But debris from his aircraft had hit and damaged Marseille's Me 109.

While Le Roux, who had been injured by shrapnel in his leg, took his Hurricane down into a crash-landing, Marseille began gliding his Messerschmitt with a stalled engine towards his own lines. And now, the incredible happened. Suddenly, Marseille spotted a lone Hurricane at a lower altitude a bit further ahead. This was flown by First Lieutenant William Herbert, a Hurricane pilot from 1 SAAF Squadron, who had had his aircraft damaged by Marseille on September 9, 1941. Marseille pushed the stick forward to gain more speed, caught up with the enemy aircraft, and fired a short but well-aimed volley of bullets. Herbert died in his Hurricane, which was engulfed in flames.

After that, Marseille steered his aircraft further and made it back to the base, where he made a nice landing. Such a long glide should theoretically be impossible—but Marseille did it.

On February 15, the two Australian fighter pilots, First Lieutenant Tommy Briggs and Sergeant First Class Frank Reid, took off with their Kittyhawks to intercept a formation of German bombers. But they did not make it very far into the air and they did not see the Messerschmitt 109 that appeared from out of the sun. Marseille picked the leading aircraft first. Briggs bailed out of his burning Kittyhawk at an altitude of only 70 meters, but his parachute barely deployed before he hit the ground. However, the pilot survived, albeit with serious injuries. Left alone against the German ace, Sergeant First Class Reid did not stand much of a chance. Marseille had pulled his '109 up in a climb towards the sun. Reid must have been terrified when his Kittyhawk took the exploding shells from the German automatic cannon. He was still seated in his aircraft as it dove steeply and crashed into a ravine.

Six days later, Homuth led a formation of six Me 109s—one of them with Marseille at the controls—on a patrol mission when they discovered a formation with eleven Kittyhawks in front of them and at a higher altitude. The British did not seem to have spotted them, so the Germans climbed straight into the sun to get above them. They were three hundred meters above their opponents when suddenly one of the Kittyhawk pilots—the ace Clive Caldwell—pulled back his stick and opened fire. The volley of bullets hit the Messerschmitt flown by Marseille's friend, Lieutenant Hans-Arnold Stahlschmidt. Homuth's stern voice was heard over the German air radio, "What kind of an idiot lets himself be shot down in this way?"[103]

After that, the '109s descended on the British unit, 112 "Shark" Squadron, which had sharks' jaws painted on the noses of their fighters. When the fighting was over, Marseille had reported two shot down and Homuth one. In reality, two Kitthawks were lost while a third was damaged.

Meanwhile, Stahlschmidt made it with a belly-landing and was soon back with his comrades. But only a few days later, bad luck struck again. On February 26, Stahlschmidt's wingman, Gerhard Keppler, returned alone from a mission with the report that Stahlschmidt had been shot down and belly-landed in hostile territory.

The next day, all of JG 27 took off to demand vengeance; the 21-year-old "Fifi" Stahlschmidt was one of the most popular ones among the unit's pilots. The German fighter pilots discovered a dozen Kittyhawks and attacked immediately. The battle raged for twenty minutes over a 30 x 40-kilometer area west and south of Tobruk. Five Kittyhawks were shot down—two from 3 RAAF and three from 450 Squadron. The

Germans reported five downings, including two by Marseille and one by Homuth.

Marseille had now reached 52 victories and was the most successful pilot of the group. Since a few days ago, he also wore the Knight's Cross. Homuth, who "only" had 42 victories himself, in spite of having been in combat service for a longer time than Marseille, was feeling increasingly uncomfortable. The enmity between the two of them deepened—"Edu" Neumann's orders that they were to fly together regularly had not had any moderating effect, rather, on the contrary.

Homuth now grounded Marseille with the motivation that he "needed to calm down." But everybody knew that it was in order to catch up with Marseilles number of downings.

Marseille had no intention of obeying this order. He started his Me 109, took off, and then returned—and carried out a low-level attack and strafed the ground in front of Homuth's tent with his machine guns!

Homuth reported Marseille for a court martial, but Neumann managed to stop the case from proceeding. He did, however, reprimand Marseille very strongly, and extended his grounding. "Unless you obey this, I cannot save you from court martial," Neumann said.[104]

It was not until April 25, 1942, that Marseille began flying again. By then, Homuth had reached 47 victories. On his first mission after the grounding, the same day, Marseille shot down two Curtiss P-40s in an encounter where the Germans claimed to have shot down ten and the Allies lost seven, to which were added four that were so badly damaged that they had to be shipped away for advanced repairs.

In April, 1942, "Edu" Neumann arranged this party to celebrate one year in Africa. Hans-Arnold Stahlschmidt can be seen at 01:57 into the clip, Marseille, Werner Schroer, and Stahlschmidt at 02:21 into the clip.

Marseille scored new "doubles" on May 10, May 13, May 16, May 19, and May 23, and on May 31, he raked in three downings. During this time, Homuth managed to accomplish one single kill. Not only can these be verified against Allied lists of losses, but a comparison with these shows that Marseille did actually shoot down more Allied aircraft in May 1942 than he claimed. On May 16, he reported two Kittyhawks as shot down, but departed before being able to see that there actually were at least three crashing.

Marseille's biggest day so far was on June 3, 1942. On that day, he was out on an escort mission for Ju 87 Stuka planes together with five other pilots from JG 27 and six from III./JG 53, a fighter wing that had recently arrived to North Africa. But eleven South African Curtiss P-40 Tomahawks, led by the group commander, Wing Commander Tristram de la Beresford, and the squadron commander, Major Jack Frost, managed to sneak up on and attack the Stuka planes without being seen, just as these recovered from their dive after the bombing. In this situation, when the dive-bomber pilots experienced a moment of blackout because of the heavy G forces, the South Africans shot down four of them. They were credited with nine air kills: Captain "Cookie" (Louis Cecil) Botha claimed three and shared a fourth one with Beresford. Second Lieutenant Cecil Golding claimed to have shot down two, one of which was shared with Captain Robert Morrison. Frost, First Lieutenant Gaymans, and First Lieutenant Vivian Muir were credited with one each.

It was a disaster for the Stuka unit, but soon afterwards, Marseille was down there, attacking the South Africans. He had spotted the Tomahawk planes first of all, and descended on them alone. In the first attack he shot down Captain Robert Pare, a veteran with sex aerial victories. Pare was killed in his burning machine. After that, Second Lieutenant Mike Martin, Cecil Golding, Robert Morrison, and Vivian Muir were all shot down in quick succession. Captain Botha's aircraft was also hit, and Wing Commander Tristram de la Beresford was wounded.[105] Marseille's wingman Rainer Pöttgen noted the times when the aircraft that Marseille had shot down hit the ground: 1222 hours,

1225, 1227, 1228, 1229, and 1233. Marseille carried out four attacks and used up exactly 360 rounds for these six downings.

In addition, another four Curtiss P-40s were reported shot down by the other German pilots, and three by Italian fighter pilots. One lost Tomahawk from 4 SAAF Squadron and three damaged British fighters are probably the ones corresponding to these claims, but the fact Marseille shot down the six pilots from 5 SAAF Squadron is quite clear: its aircraft crashed a couple of dozens of kilometers west of Bir Hacheim, while other downings were reported thirty kilometers further east.[106]

At 01:04 in this film clip, Marseille's machine gun cameras show three of his six kills on June 6, 1942.

Cecil Golding survived and recounted after the war how it all happened when he was shot down by Marseille on this day.

Marseille was now, with 75 aerial victories, the most successful ace of the group. He had left his squadron commander Homuth with 48, far behind him. On June 6, Marseille was awarded the Oak Leaves for the Knight's Cross. Soon afterwards, there was a regrouping of the command of JG 27 that Homuth was surely grateful for: "Edu" Neumann was appointed to command the entire JG 27, Homuth was given his place as group commander at I./JG 27, and Marseille became squadron commander of 3./JG 27. On June 7, the same day as this took place, Marseille saved Homuth by shooting down two Tomahawks that were on his tail.

Of course, the opposing side knew about Marseille by now. There, they were starting to become worried about him, and tried to organize for him to be shot down. An order of the day from the headquarters of the British Middle East Air Force in June 1942, reads, "Marseille is the best the Germans have here. Like the others, he flies a Bf 109, but he flies it better than all the rest. He can only be attacked by several planes at the same time. You must make sure to attack him from the front or flanks before he is in a position to maneuver."

But none of these attempts succeeded; instead, Marseille shot down even more Allied fighters, perhaps as a consequence of the Allied fighter pilots now actively seeking to engage him. During the next few days, he bagged twenty-four Allied fighters on six missions. On June 10, there were four in a single fight; this was repeated on June 15 and 16. How well these claims correspond with reality is apparent from the Axis reporting 15 Allied fighter planes being shot down on the latter day, while the true Allied losses were 19 fighters.

On June 17, Marseille experienced his toughest aerial fight so far. By midday, he was out on a fighter sweep together with his Schwarm (four aircraft) when they suddenly came across twenty Curtiss P-40s and twelve Hurricane fighters, led by Squadron Leader Derek Ward. The 24-year-old New Zealander Ward was one of the most experienced Allied fighter pilots in North Africa. He had been in combat ever since the Battle of France in May 1940, where he scored his first aerial victories, and took part in the Battle of Britain. By now, he had 7½ confirmed aerial victories.

Without hesitating, Marseille dove straight into the mass of Allied fighters and immediately shot down one of them, but then found himself being attacked by several of the others. While banking sharply so as to shake them off, he spotted a British fighter in front of him for a split second. Marseille fired instinctively, and this aircraft, too, caught fire and crashed.

When he noticed that he had gotten rid of his pursuers, he saw four Hurricanes circling a pilot dangling from a parachute, as if to protect him. The British combat report shows that it was Squadron Leader Ward who had turned to protect his two comrades who had bailed out. Marseille descended on them and shot down Pilot Officer Gerald Wooley, who did not stand a chance of getting out of his Hurricane at such a low altitude. But Ward skillfully avoided Marseille's first attack. Marseille had shot down three aircraft — three Hurricanes from 73 Squadron—in three minutes, but now, a hellish dogfight between the two aces began. The minutes went by. Marseille was sweating profusely, and several times, he was close to losing consciousness because of the G forces produced by the narrow turns. In fact, the Hurricane was more agile than the Me 109 F, but eventually, Marseille had succeeded in out-turning Ward. A short hail of bullets and Derek Ward crashed to his death.

By now, Marseille was quite exhausted, and actually wanted to abort the fight, but the last Hurricane had been his 99th aerial victory. His headphones were ringing with congratulations from his comrades, and exhortations, "And now number one hundred, Jochen!"

Marseille could not leave. In front of him, he saw a lone Curtiss P-40 at an altitude of only one hundred meters, and attacked.

Seven pilots from 112 "Shark" Squadron had been tasked with carrying out a fighter-bomber mission, but Flight Sergeant Roy Drew had been having problems getting his Kittyhawk off the ground, so when he finally came up into the air, the others had already left. The 27-year-old "Rick" Drew had, after arriving at 112 Squadron in October, 1941, made a name for himself as a "courageous and determined pilot" who always was "a role model to others."[107] With 70 combat missions in his logbook, he was very experienced. During these missions, he had on one occasion shot down two Italian fighters in combat, and had been shot down himself once and had had his aircraft badly damaged in combat on another occasion, so he had no intention to abort but set his course for the front on his own. Marseille shot him down already in his first attack.

When Drew plummeted to his death, Marseille had scored his one hundredth aerial victory. This seems to have invigorated him; when he saw two Spitfire plans high above him, he climbed towards them. Flight Lieutenant Francis Spicer does not seem to have discovered the danger—it is often difficult to spot aircraft painted in camouflage colors against the ground below—before his Spitfire was shaking from the hits from Marseille's guns. The Spitfire went down in flames and slammed into the ground, marking Marseille's 101th aerial victory.

Now, Marseille felt that he had had enough, and set course for home, but he had barely done so before being caught up by a formation of five Hurricanes. Those were the ones that remained airborne from 73 Squadron. Luckily enough for Marseille, the British pilots did not discover that it was a German Messerschmitt that they had caught up with, but for five minutes, Marseille continued to calmly fly in formation with them, while he regained his strength. Then, he decided to unmask himself and went into position behind one of the British fighters and pressed the fire buttons. Nothing happened! All three weapons jammed at the same time! There was nothing else to do, so Marseille quickly dove away from there. But two Hurricanes detached from the British formation, and went after his Me 109. These, too, were flown by anything but beginners: Flight Lieutenant Ivor James Badger and Pilot Officer Herbert William Coussens were both veterans from the Battle of Britain. Moreover, Badger, who had been a fighter pilot since 1938, had also belonged to a group that was specialized in advanced flying.

Marseille literally had to run for his life. He descended to the lowest possible altitude with Badger and Coussens on his tail. All the time, he had to throw his aircraft to one side after the other to avoid the British fire. These British pilots were flying the new Hurricane version, Mark IIc, which was equipped with four 20 mm automatic cannon with a combined rate of fire of 40 shells per minute. Thus, it would be enough with one split-second hit to tear the 109 to pieces. Eventually, Marseille managed to turn his aircraft sharply in a breakneck maneuver. He flashed past the two Britons, who were completely taken by surprise, and disappeared at full speed.

As Marseille landed a few minutes later at the airfield, there was a jubilant reception committee awaiting him. But "Jochen" remained seated in the cockpit, his arms hanging loosely, and staring emptily. The fight had taken his last strength.

The next day, Marseille stepped aboard a trimotor Junkers 52, which was to take him to Germany. He was going to meet Hitler to receive the Swords to the The Knight's Cross with Oak Leaves, and, in connection with this, he would be used by the German propaganda during a few weeks. Hans-Joachim Marseille was now one of Germany's greatest war heroes. He was dubbed "the Star of Africa" and his image was in newspapers and on posters all over the country.

But "Terrible Jochen" would not become what the Nazi leaders had expected.

Marseille visits Field Marshal Rommel in his personal Kübelwagen called "Otto."

Hitler's personal pilot, Hans Baur, recounted how things began with dinner with Hitler, Göring, and the Minister for Propaganda, Goebbels, after the award ceremony. "How does it feel to have had your one hundredth?" Göring asked, upon which Marseille immediately responded, to the horror of the assembled Nazi leaders, "Do you mean aircraft or women?"[108] Afterwards, Hitler's personal air force adjutant, Colonel Nicolaus von Below, asked whether Marseille had considered joining the Nazi Party, upon which Marseille replied, "If I ever find a party worth joining, I will join it."[109] Bauer recounted after the war that "Marseille made a few most unflattering remarks about the Nazi Party."[110]

The Hitlerjugend leader Artur Axmann wanted Marseille to hold a patriotic series of lectures for Hitlerjugend youth, but concluded that Marseille was "the ultimate role model for German youth—until he opened his mouth." Axmann also said, "Without doubt, Marseille was the least soldierly man I have ever met. He seemed so incredibly out of place with other soldiers. He did not act like an officer at all, rather like a schoolboy."[111]

During a reception with the entire Nazi elite—including Himmler, Goebbels, Bormann, Göring, Axmann, Wolff, and Hitler—Joseph Goebbels' wife Magda said that she had heard that Marseille was a talented piano player and asked him to play a piece. There was a large grand piano in the room, and Marseille went and sat behind it. The audience turned politely towards him and nodded with approval as he began playing Beethoven, followed by other classical pieces. But suddenly, there was a slight smile on Marseille's lips, and the next moment, he segued into a jazz tune at breakneck speed—"The Entertainer" by the African-American musician Scott Joplin.

The Entertainer with Scott Joplin.

Playing this in front of the top Nazi elite was something absolutely outrageous. The Nazi regime, which was as conservative as it was racist, had outlawed all music performed by African-Americans, which they called "degenerate music" and "negro music." They had even been arranging exhibitions to warn the German public of "degenerate music," where they claimed that the Jews used people of African descent to ruin the German culture by "negro music." There was an underground German youth group in Berlin which called itself *Swingjugend* and which convened in secret to listen to and dance to jazz music under circumstances characterized by a generally "free" way of living. They were decidedly anti-Nazi and used to greet each other with the words "Swing Heil" (instead of "Sieg Heil"). The German Reich Ministry of Justice wrote in a report on them, "One of the most prominent among these dangerous groups in Germany is the so-called *Swingjugend*. Their heretically-minded ideas about individual freedom are leading them into openly opposing the Hitlerjugend. Their meetings are characterized by drunkenness free sexuality, and dancing orgies, where the teenagers are 'swinging' and becoming 'hot.'" It is quite possible that the Berliner Marseille was in contact with these. In that case, perhaps his animosity towards the Nazi leaders to a certain extent could be explained by several of the leading people among the *Swingjugend* having been arrested and thrown into concentration camps. Perchance the jazz melody he played for the top brass was his tribute to them?

Hans Baur recounted, "I was looking at Hitler and the others. They could not have looked more horrified, even if Winston Churchill himself had stepped inside with a pistol in his hand."[112] Artur Axmann recounted that he was gripped by such fear that it felt as though his blood froze. Hitler rose, furious, saying, "I think we've heard enough," and marched out, immediately followed by Göring and Goebbels. One after the other, the Nazi leaders drifted away. "Marseille's jazz melody only lasted for a few minutes, but it efficiently emptied the entire room of people," Baur remembered.

Had he not been such a war hero, Marseille could have landed in a concentration camp for this, but now, he escaped without punishment. But this gathering would change much for him in another way. He overheard a conversation between two high SS officers, Odilo Globocnik and Karl Wolff. Globocnik was responsible for Operation "Reinhard," the Holocaust plan that led to the murder of about one million Polish Jews in gas chambers, and it was this that they were talking about when Marseille happened to hear what they said.

What happened now has not been entirely clarified, since it seems that there are no documents about it and the main person is dead, but Marseille clearly changed. In early August, his leave was over, and he was to return to JG 27 in North Africa via Italy. But first, he was to go to Rome to receive the highest Italian decoration for valor, the Gold Medal, from Mussolini. That happened on August 13, 1942, and after that, Marseille was due to board a Junkers 52 which was to take him back to his unit, which now was stationed in Egypt.

But Marseille never showed up at the airport, so the transport aircraft flew to North Africa without him. When the military police failed to find him, the Gestapo was called in. The hunt for Marseille was led by the infamous SS-Obersturmbannführer Herbert Kappler, who eventually managed to capture the airman.

The military historian Colin D. Heaton and his colleague Anne-Marie Lewis write in their biography of Marseille: "Somehow, they managed to convince him to return to his unit. It is remarkable that there were no notices made of this in the formal report."[113] Franz Kurowski does not mention anything about this in his otherwise quite detailed Marseille biography. But there is a story circulating among the veterans that is supported by certain circumstantial evidence: Because of what he had learned about Operation "Reinhard," Marseille had tried to defect and flee to Switzerland, but he was arrested by the Gestapo, who gave him two options: either he would return to his unit and continue living up to the role as the National Hero, or both he and his beloved mother would be thrown into a concentration camp. This ultimatum was allegedly followed by a warning—that he needed to prove himself worthy of his and his mother's freedom by continuing to shoot down enemy aircraft at the same pace—and a threat: if he would allow himself to be captured, his mother would be put in a concentration camp.

It is difficult to assess to which extent this is true. What we do know is that when Marseille returned to North Africa on August 23, 1942— ten days late—his comrades found him a changed man. He was grim and seemed depressed. And he did not want to say why he had gone AWOL, which definitely was unlike him. Had it been because of an adventure with a woman, which was the most common reason for him going AWOL, he would probably—as so many times before—have happily been bragging about this. It was common knowledge that he could not keep quiet about such things. But he unburdened his mind to at least one person, telling him what he had heard about the mass murder of Jews, which had affected Marseille very seriously. "He had started drinking again. It was obvious that he was not the same person as when he had gone on leave."[114]

If the Gestapo and the Sicherheitsdienst (SD) had demanded that he would perform as before in aerial combat, they would not be disappointed. Suddenly, Marseille fought like a man possessed, shooting down enemy aircraft at breakneck speed.

By this time, Rommel had driven back the Allies from Libya and was now facing el-Alamein in Egypt. Both sides tried to reach a decision here through an extensive effort with air power, artillery, and tanks. But during the first few days after Marseille had returned, things were calm in the air. He flew mission after mission without seeing any Allied aircraft. One remark that he let out during one of those days is quite remarkable, "If nothing happens soon, I'll be forced to shoot myself down."[115] His comrades took it as a joke, but there could have been more to this remark than they knew.

But on August 31, 1942, the calm was over as Rommel began a new attempt to break through the British positions. Marseille flew escort missions twice that day for large formations of Ju 87 Stukas, with a total of sixty dive-bombers. He said that he liked this because the Junkers planes acted as "magnets on the opponent's fighters." On the first occasion, he shot down two Hurricanes from 213 Squadron: both pilots made it. On the second occasion, he and his comrades clashed with a large number of Spitfire, Hurricane, and Kittyhawk fighters. Marseille began by shooting down the Spitfire pilot Sergeant First Class Kenneth Lusty, who made it by parachute.

The next day, September 1, 1942, became the day that Marseille has become most famous for. It was then that he broke all records by shooting down no less than 17 Allied fighters in one single day. These were distributed over three aerial clashes: between 0826 and 0839 hours, he shot down four, between 1055 and 1105, eight, and between 1747 and 1753, five. This has been questioned by many, not least on the British side. On this day, so many aircraft were up in the air from both sides, and the exact times are not always correct, that it is difficult to determine to what extent Marseille's claims were in line with reality. But the Allied side reported the loss of a total of 26 aircraft over el-Alamein on that day, while the Germans reported as many shot down (26, 17 of which by Marseille) and the Italians nine.[116] What is obvious is that the British lost at least seven fighters in Marseille's last aerial battle on that day, while Marseille claimed to have shot down five and his comrades another four.

A remarkably large number of the pilots who were shot down by Marseille made it either by bailing out by parachute of by belly-landing. On September 2, he shot down four Curtiss P-40s and one Spitfire— all pilots survived. On average, the chance of surviving being knocked down was otherwise at about 50 %.

The next day, when Montgomery's Eighth Army repulsed Rommel's attack, a great aerial battle once again raged over el-Alamein. The German fighter pilots reported 18 victories without any own losses, and the Italian fighter pilots seven, without any own losses as well. On the Allied side, fifteen fighters and one bomber were lost. Marseille contributed to the German total of six kills. On that day he was awarded Germany's highest military award, the Diamonds for the Knight's Cross with Oak Leaves and Swords, as the twelfth soldier in Germany.

Marseille followed it up with four kills in one and the same aerial battle both on September 5 and 6. His claims still agreed very well with real Allied losses; out of a total of sixteen Axis claims on September 6, fifteen can be corroborated by Allied loss reports. The next day, I./JG 27 was credited with the shooting down of four Curtiss P-40s, including two by Marseille, near el-Alamein; South African 4 and 5 SAAF squadrons lost four Curtiss P-40s in that battle. Once again, all pilots made it, either by bailing out or by crash-landing without any German aircraft (as was customary) coming down and to shoot their planes to pieces, upon which the pilot was often killed or wounded.[117]

After a few days' suspension in air activities because of a sandstorm, Marseille was in the air again on September 11, searching for new enemy aircraft to shoot down. But this time, things did not go as well. Together with twelve Italian MC 202 Folgore fighters, he and eight other Me 109 pilots attacked some twenty British aircraft—Hurricane fighter-bombers escorted by twelve Spitfires. Marseille was credited with the shooting down of two "Curtiss," but a comparison with British combat reports shows that Marseille at most could have damaged a Hurricane.[118]

Kommodore "Edu" Neumann noticed that something was wrong with Marseille—he was tired and seemed worn—and therefore, he grounded him for a few days, which Marseille objected violently against. When he was allowed to fly again, on September 15, he claimed to have shot down seven Curtiss in a fight where other German fighter pilots reported 13 kills. This time, the German fighter pilots exaggerated their figures considerably: The Allied losses were limited to six destroyed and two damaged aircraft.

Once again, Marseille was grounded—now for ten days, on the pretext that he had been promoted to Captain, which was a bit odd. But "Edu" Neumann disclosed his true reasons for this afterwards: he had noticed that "Jochen was completely exhausted, physically and psychologically."[119]

Marseille attempted to circumvent his grounding by visiting the Italian fighter wing 4. Stormo a few days later. He asked to "try out" a Folgore fighter. The group commander, Lieutenant Colonel Armando François—himself an ace with ten aerial victories—knew nothing about the grounding, and was only happy to meet the famous German's wishes.

There are a few photographs of Marseille during his visit to 4. Stormo. He is looking quite depressed in the images, which were taken before the flight. As he then took off with the the Folgore, things went badly. If it was down to Marseille's stress level or that he was unfamiliar with the aircraft (which was, however, quite reminiscent of the Me 109) is not known, but suddenly, he accidentally turned off the engine, and only barely managed to belly-land the Folgore, which was badly damaged.

On September 25, Neumann informed Marseille that his grounding was over, effective the following day. Marseille eagerly went to find his wingman, Joseph "Jost" Schlang. "We'll take off as early as possible," he said. "Tomorrow is a new day for shooting down!"

Marseille was up in the air with his Schwarm early in the morning of September 26. He attacked a formation of Hurricanes and shot down Pilot Officer Clifford Luxton, who made it in a crash-landing. After that, the Me 109s were attacked by eight Spitfires from 92 Squadron, which had been flying as top cover for the Hurricanes. In the ensuing dogfight, Marseille was claimed to have shot down three and "Jost" Schlang one of the Spitfires. 92 Squadron's actual losses were confined to two Spitfires. Pilot Officer David Turvey, who managed to bail out, probably became the last pilot to be shot down by Marseille.

That afternoon, Marseille and eight other Me 109 pilots were out on an escort mission when the six Spitfires appeared at a lower altitude. These came from 145 Squadron and attempted to sneak up on the Stuka planes from below, but Marseille immediately descended on them.

But once again, things did not work out for him. After attacking again and again, he suddenly ended up in a duel with a Spitfire flown by Flight Sergeant Francis Barker. The 21-year-old Australian Barker was no particular ace; he had been serving with 145 Squadron since the end of 1941 and had been credited with the destruction of two Me 109s, of which at least one is unverified. But now, Barker, with his Spitfire Vb,"DL-M," engaged North Africa's probably best fighter pilot in an eleven-minute dogfight, which ended with Marseille steering straight into the sun with Barker on his tail. When Barker returned to base, he was able to report a damaged Me 109, while Marseille was credited with three kills—Nos. 156-158. In reality, no aircraft was lost on either side, and none of the British fighters was even hit by fire.[120]

Based on eyewitnesses, Franz Kurowski describes the scene as Marseille had landed on the airbase: "As he climbed out, his two crewmen were horrified. They had never seen their Jochen looking like this. He appeared to be a ghost of their Staffelkapitän. Trembling hands grabbed for a cigarette which Meyer passed to him. He took one, two deep drags. His face looked like that of an old man.

He went to the flight debriefing in a trance. There he confessed: 'a mastery adversary. Never before had an enemy fought like he did. I don't know how things will turn out next time.'"[121]

But it was Marseille himself who could not take any more. Francis Barker was credited with yet another 1½ aerial victories (on October 21, 1942, and March 29, 1943) before he was killed in an air accident on August 7, 1944. Marseille had met and defeated many pilots who were considerably more skilled and experienced than him.

All of this was unlike Hans-Joachim Marseille. Having previously seemed to have limitless energy, he had now been completely exhausted in only a few weeks after having returned from two months' leave. He had also started exaggerating his victory results in a completely different way than before and seemed unfocused—he was definitely not as on edge as before—in aerial combat. It is undeniably easy to get the impression that there was something eating him.

Marseille surprised everyone when Field Marshal Rommel called him personally on September 28 and offered Marseille to accompany him to Berlin, where he would be able to meet his beloved mother again—and Marseille declined. Turning down a field marshal, especially the "Desert Fox" Rommel, was actually an act of insolence if you were a mere 22-year-old captain. "Don't be silly, Marseille," Rommel replied in surprise, "pull yourself together and fly to Berlin together with me!" But Marseille stood his ground, and so did Rommel. "At least, you ought to think about it," the latter insisted. Eventually, Marseille burst out pleading, "I beg you, *Herr Feldmarschall*, not to hold on to this invitation!"—"Well, stay then, for God's sake!" Rommel replied.[122]

No-one could understand why Marseille, who was completely exhausted, went to such lengths to remain at the front.

Now, a few "calm" days followed. September 30, 1942, was also one of those days with low activity in the air. At 1047 hours, 23 Me 109 took off from JG 27 to escort Stukas against targets behind the British lines at el-Alamein. Marseille led the formation. It was his 388the sortie. This time, he flew one of the brand-new Me 109 Gs, version "Gustav." A few of these had recently been delivered to JG 27, but Marseille had refused to fly them because he did not trust their new engines. For some reason, however, Field Marshal Albert Kesselring, the highest Luftwaffe officer in the Mediterranean area, had personally ordered Marseille to fly an Me 109 G. Therefore, he was now seated in the cockpit of Me 109 G-2, serial number 14526, with a type-605 A-1 Daimler-Benz engine, serial number 77 411.

Neither on this occasion did the Germans encounter any hostile aircraft in the air. When the Stuka planes had carried out their attack, the formations returned towards their own lines. "Edu" Neumann sat at the base, following the radio communication between the airmen. At 1130 hours, a high-pitched voice was suddenly heard: "My engine's on fire!" It was Marseille who had made the call, and he continued, "I've got a heavy build-up of smoke in my cockpit. I can't see outside anymore!"

Something had gone wrong with the engine, which emitted ever-thickening smoke. The aircraft lost more and more altitude. But Marseille continued flying. "Are we over our own lines yet? I can't see anything!" he cried desperately over the radio.—"Two more minutes, Jochen!" his wingman, Rainer Pöttgen, replied. At the base, Neumann was beside himself with apprehension. "Bail out!" he shouted over the radio, but Marseille refused, in spite of flames now shooting out from the engine. It must have been unbearably hot inside his cockpit. The seconds were dragging on and became minutes. The increasingly burning 109 descended more and more. Eventually, the call came over the radio, "We're over el-Alamein now!" Soon afterwards, Neumann heard Pöttgen's voice over the radio, "Jochen has jumped!" Then, there was a moment of silence, and then, Pöttgen's report in a broken voice, "He's dead!"[123]

Marseille had turned his Me 109 G on its back, shot the canopy off, bailed out, and hit the tailfin. Apparently unconscious from the hard collision, he plummeted to his death.

Hans-Joachim Marseille as a young and cheeky fighter pilot on the English Channel in 1940. (Photo: Steinhoff.)

Marseille became notorious for his many romantic escapades that often had a negative effect on his military duty. (Photo via Neumann.)

Eduard "Edu" Neumann. (Photo: Neumann.)

Johannes Steinhoff was a very demanding commander. He was in active front service between 1939 and 1945, flew 993 combat missions and was credited with 178 victories. (Photo: Steinhoff.)

On April 23, 1941, Marseille was outmaneuveredand shot down by the French Sub-Lieutenant James Denis. (Photo: Neumann.)

Marseille and a local resident inspect a Hurricane he just shot down. (Photo: Neumann.)

Marseille in jersey day. (Photo: Neumann.)

The contradiction between Marseille (r.) and his division manager Gerhard Homuth (l.) is quite evident in this picture. (Photo: Neumann.)

The pilots of 112 "Shark" Squadron, RAF, in front of one of the unit's Kittyhawk fighters in January 1942. From the left: Squadron leader Clive Caldwell, Sergeant William Carson, Sergeant Andy Taylor (on the wing), Flight Lieutenant Neville Duke, Flight Lieutenant Peter "Hunk" Humphreys, Henry George Burney (sitting on the wing with his hand on the tailpipe), Pilot Officer Roy Drew with his arm on Sergeant Alexander Donkin. The man behind Donkin is unidentified. To the right of Donkin, dressed in a white overall, Sergeant Rudolf Leu, and far right Flight Lieutenant Eric Dickinson.

Roy Drew became Marseille's 100th Air Victory on June 17, 1942. Donkin also was killed when he was shot down by Marseille on February 8, 1942, and Marseille probably shot down William Carson's brother, Kenneth Carson (also in 112 Squadron) on June 6, 1942, whereby Kenneth Carson ended up in Italian captivity.

Hans-Joachim Marseille and his mother Charlotte Marseille.

A much more serious Marseille, with the Knight Cross with Oak Leanes, Swords and Diamonds, in the summer of 1942.

Marseille on the cover of the German flight magazine Der Adler, the Spanish edition.

Curtiss P-40 was Marseille's most common opponent in air combat. In total, he claimed exactly 100 aircraft of this type shot down. (Photo: NARA.)

CHAPTER 5

'Let's Not Talk About That' — Uncomfortable Facts

"But let's not talk about that." Hearing Luftwaffe veterans saying that during my interviews and conversations with them is nothing unusual. Many times, it has been preceded by a portion of ordinary gossip about other airmen. But sometimes, it is about of more serious matters, and sometimes, it is about things that I could not reproduce in a book without risking being sued by next of kin. It could be about alleged murders of other airmen during the war (on one occasion, two German fighter aces—each with more than 100 aerial victories—were supposed to have been dueling using fighter planes, with the consequence that one was killed and the other was wounded; the official cause of death for the one who was killed was that he was shot down by American fighters, but on that date, there were no American fighter operations in that area), it could be about suicide, about drunken deaths after the war, or about massively exaggerated claims of air victories (a comparison with the opposing side's loss reports shows that some aces with officially more than 200 aerial victories in reality shot down 20-40 aircraft). It could also be about what was perhaps the most sensitive thing of all, collaboration with the enemy. "If he had been alive today, he would have sued you," one very irate German aviation historian wrote to me after reading in one of my books that Russian documents showed that one downed and captured Luftwaffe pilot had been collaborating with the Russians.

But some pilots happily admit to having agreed to collaborate with the enemy. Perhaps the most well-known one of them all is the fighter pilot Heinrich Graf von Einsiedel. Born in 1921, von Einsiedel was truly blue-blooded. The "Iron Chancellor" Otto von Bismarck was his maternal great-grandfather. His father was a Prussian officer and his sister Gisela married Wolfgang von Richthofen, a younger relative of the "Red Baron." During the 1930s, the young Heinrich was the leader of an underground anti-Nazi youth organization—*Swingjugend* (see the chapter on Marseille)—but was caught, arrested, and subject to many strict interrogations. He managed to make it out of there because of his good connections and by volunteering for the Wehrmacht, having barely turned 18.

With an acquaintance as an intermediary, he managed to get admitted to the Luftwaffe, arriving as a freshly-graduated pilot and cadet at JG 2 "Richthofen," where he got to know "Assi" Hahn. Von Einsiedel turned out to be a very talented pilot. His first downings were two Swordfish torpedo planes as he was covering the *Scharnhorst's and Gneisenau's* break-out from the English Channel on February 12, 1942. Soon afterwards, he was transferred to the fighter group III./JG 3 "Udet" on the Eastern Front.

Von Einsiedel recounted that he, in those days, "saw the war as an adventure, as a way to gain fame and medals. I didn't think that much about how the war was proceeding or about the Hitler system. We were taking part in a 'Crusade on Bolshevism' without knowing much about what that actually was."[124] But at III./JG 3 "Udet," he found himself in the same position as Marseille—in conflict with a commander who also was displaying envy towards von Einsiedel's success. The group commander Captain Wolfgang Ewald, ten years his senior, was everything the von Einsiedel was not—a strict and demanding, convinced Nazi. The short-statured Ewald had been flying as a volunteer in the Spanish Civil War, where he had been credited with one aerial victory, and had participated in the Battle of Britain.

Ewald found it hard to accept that the long-haired and quite easy-going "puppy" von Einsiedel surpassed him in aerial combat. Ewald had 15 aerial victories when von Einsiedel turned up. After one month, von Einsiedel had 12 and Ewald 18. During August 1942, Ewald had to see himself completely cut out by the "upstart": on August 6, von Einsiedel shot down four Soviet bombers and increased his tally to 16. Three days later, he caught up with Ewald by reaching his 20th victory. On August 12, von Einsiedel surpassed his commander by bagging his 21st and 22nd. Ewald was credited with his 21st on August 13, but meanwhile, von Einsiedel brought home his 23rd, followed by the 24th and 25th during the next few days. Ewald was happy to shoot down his 22nd on August 19, but the next day, von Einsiedel superceded his personal record by shooting down five Soviet aircraft—his victories Nos. 27-31—which was more than Ewald had managed to accomplish in four weeks.

The conflict between the two of them was intensified as von Einsiedel humiliated the veteran and unit commander Ewald in front of the entire unit. But on August 30, 1942—when von Einsiedel's tally contained 34 aerial victories (won during 170 missions) and Ewald had 24—the competition was over. Von Einsiedel was shot down by a Soviet fighter pilot and ended up in captivity.

Excerpt from *Black Cross/Red Star: Air War over the Eastern Front, Volume 3* by Christer Bergström:

"Von Einsiedel recalls his last mission: 'Only ten or twenty miles away, on the other bank of the Volga, lay the Russian airfields. The Russian fighters were already piercing the blanket of clouds in a dive attack on us. One of them seemed to have lost his comrades and arrived right in front of us. Major [Wolfgang] Ewald needed only to turn in slightly to get into position to shoot. He fired, and a white trail behind the Russian showed that his engine had been hit. But Ewald was already beyond the LaGG so I turned my machine to complete the job with my three guns. The Major yelled into the radio, "Leave him to me, leave him to me!" "After you, Herr Major," I replied. There was irony in this because it was not customary for us to address each other by rank when in battle. But this was not the first time that the Major had played this game with me, and it was not the first time that he was too late. Before he was able to get into a firing position again, the Russian had made a nice belly landing and the clouds of antiaircraft fire around us showed that he had not chosen a bad place to land. But Ewald had another idea: "Einsiedel, you have three guns, set him on fire!" The belly landing of the LaGG on his own territory would not count for the Major unless the machine were destroyed on the ground. Without his order I was about to attack the plane on the ground. But in my anger over such unnecessary experiments fifty miles behind the front I could not resist asking once more, 'Wouldn't you rather yourself, sir…'

Then I fired. In the fire of my guns the LaGG broke up on the ground and went up in flames.'

Next, Major Ewald's pilots came across a mixed 102 IAD-PVO formation--I-16s from 629 IAP and I-153 biplanes from 926 IAP--on the south side of Stalingrad, where the Volga bends sharply toward the southeast. Although hesitating to engage 'these troublesome insects … too agile for our fast machines,' as von Einsiedel expressed it, the III./JG 3 pilots turned down against the Polikarpovs. The Soviet formation leader directed his Polikarpov fighters into a loop over the diving Bf 109s, then they turned their little planes almost on the spot and came at the Germans head on with blazing guns. The two I-16 pilots Leytenant Fyodor Fyodorov and Starshina L. I. Yershov both scored hits on von Einsiedel's Bf 109. With liquid streaming out of his aircraft's right radiator, von Einsiedel started gliding toward the front. He could just get over the Zarepta hills. Then, 'with a sinister jerk the propeller stopped.'

Right in front of him, von Einsiedel saw a Soviet airfield. Minutes later, the dazed great grandson of Otto von Bismarck climbed out of his belly-landed Bf 109 on the runway of this VVS airstrip. What then happened stunned von Einsiedel, who had feared the worst:

'A car approached and an officer in pilot's fur boots got out. He held out his hand. A little gold star glittered on his chest. "Comrade," he said, rolling his r's. At last I could put down my hands. Greatly relieved. I took the hand he offered me.

'With the air of an expert he examined the wreck of my machine lying forlornly on the sand, with its buckled wings, bent body, and twisted propeller blades. When he saw the little white lines on the controls, and the cockades and Soviet stars, he gave an exclamation of astonishment. He counted them quickly – ten, twenty, thirty, thirty-five – then nodded his head thoughtfully, and forming the figure twenty-two with his fingers pointed to the gold star on his chest and said something in Russian. As I did not understand, he added, "I – Hero of Soviet Union." Not realizing that this was an official decoration given for special merits, I could hardly suppress a smile. But he seemed to take it for a sign of recognition, and laughed with pleasure and pride."'

There is only one Soviet pilot in the Stalingrad sector who fits with von Einsiedel's description, and that is the 21-year-old Captain Michail Baranov. He had 22 aerial victories at this time—and was the most successful Soviet ace in the Stalingrad sector—and had been appointed Hero of the Soviet Union on August 12, 1942.

Arkadiy Kovachevich, who served as a Starshiy Leytenant with 27th IAP on the same airfield, still has a vivid memory of the meeting between the Soviet fighter pilots and the captured Leutnant von Einsiedel. According to Kovachevich, von Einsiedel spent five days with the VVS pilots on the air base. Von Einsiedel describes how he was beset by a hailstorm of questions from several Soviet fighter pilots: "Why did I have to make a forced landing? How many planes had I shot down? What decorations had I received? Was I married and where was my home? All these questions were put in broken German with laughter, astonishment, and childlike curiosity." Due to von Einsiedel, the treatment that he received pendulated between a colleagual amiability and hostility, including physical violence. However, the "group captain" who had first received him, interfered against a furious Soviet pilot who had assaulted von Einsiedel physically, giving the attacker "a dressing down of the kind one used to hear on a Prussian barrack square," according to von Einsiedel."

In captivity, von Einsiedel was—as were many other German soldiers and airmen—subject to conversion attempts, with good success. The Soviet interrogators were experts at this. It went as follows: the chief interrogator, Colonel Sergey Tulpanov, spoke fluent German. He immediately offered von Einsiedel to forward a letter to his family. Von Einsiedel happily wrote such a letter, where he described how well he was being treated, and, to flatter his captors, he added, "I'm convinced that we'll never be able to defeat such a great state as Russia. We should end the war as quickly as possible." A few days later—on September 3, 1942—a Soviet aircraft dropped this letter, reedited into a flyer, aimed at "German airmen, the officers and comrades, dear comrades at fighter wing Udet," in thousands of copies over the German lines and JG 3's airfield.

With that, all bridges were burned for von Einsiedel, and soon, the conversion attempts started taking effect. In July 1943, he took part in founding *Nationalkomitee Freies Deutschland* (the National Committee for a Free Germany), the German organization for prisoners of war who took a stand against Nazism. There, von Einsiedel was a member of the board.

The previously frosty relation between von Einsiedel and Ewald switched into direct hostility after Ewald had also been shot down and ended up in Soviet captivity. Ironically enough, this took place on July 14, 1943, the day after von Einsiedel had taken part in founding the National Committee. Von Einsiedel and Ewald met again in captivity, but now, the roles were reversed. Ewald belonged to the "refusers' front" among German prisoners of war, who strongly rejected the National Committee. Later during the war, von Einsiedel spread propaganda over loudspeakers at the front to the German lines, intending to make German soldiers desert.

But the Soviet system meant a great disappointment for von Einsiedel. He was terribly upset by the traces of Soviet abuse of German civilians that he saw when following the Red Army's advance. When the war ended, the Russians dissolved the National Committee and kept von Einsiedel as a prisoner of war. When he finally, after five years' imprisonment, was released, he was initially only allowed to settle in East Germany.

When he was then allowed to travel to West Germany to visit relatives in May 1948, he was arrested and jailed by the American military authorities in Wiesbaden. He was formally accused of having travelled on a false visa, which was not true. After a hunger strike, he was released on bail in October 1948. During the trial in November 1948, he was acquitted on all charges, but was still deported to East Germany.

But he was being watched by the authorities in East Germany too. He recounted, "I couldn't live in the Eastern Zone, because my notion of Socialism was something completely different than Stalinism."[125] As early as in December 1948, he moved to West Berlin. There, he started writing his memoirs (*Tagebuch der Versuchung: 1942-1950*, published in English as *I Joined the Russians: A Captured German Flier's Diary of the Communist Temptation*) in which he openly recounted his collaboration with the Soviets during the war. These were published in West Germany (they were banned in the East) in 1951 and caused a storm of indignation. One of his strongest critics was his former comrade "Assi" Hahn, who had been very badly treated in Soviet captivity.

The victim of persecution in both parts of Germany, von Einsiedel applied for a visa to emigrate to the USA but was denied. In 1957, he decided to join the West German Social Democratic Party, where he was accepted. But when he visited a veterans' meeting for former airmen and ground staff from JG 3, he was ostracized and had to leave the site early. "Those people hadn't learned anything," von Einsiedel said bitterly.[126]

In 1993, he left the Social Democrats as a protest against what he felt was "dirty compromises" and joined, together with a few other disappointed Social Democrats, the new Party of Democratic Socialism (PDS). In the election of 1994, von Einsiedel was elected to the German Bundestag as its oldest member. He died in Munich in 2007. The fate of Heinrich Graf von Einsiedels reflects the climate for Luftwaffe veterans in Germany during the post-war period quite well.

Another Luftwaffe airman from the high nobility, who switched sides and became a politician after the war, was Egbert von Frankenberg und Proschlitz. He belonged to the so-called ancient military nobility going back to the Middle Ages. His family tree is full of knights, commanders, and generals. He joined the Nazi Party as early as in 1931 and the SS in 1932. But at the same time, he trained to become an aviator. He volunteered for the Condor Legion and flew bombing missions during the Spanish Civil War.

But a series of very serious losses in his unit would affect him strongly. In the fall of 1942, he served as commander of the bomber group III./KG 77, which was deployed in a last desperate air offensive on Malta. By then, the island's air defense had been reinforced considerably. Von Frankenberg's group had 27 Ju 88 bombers on October 1, 1942. Of these, 24 were lost before the month was over. But with new additions of both aircraft and crews, the air offensive was ordered to continue, to the horror of von Frankenberg. Day after day, he saw his young crews disappear over the Mediterranean. In November, another 14 aircraft were lost. Eventually, the unit was taken out of service altogether. The remaining aircraft were handed over to the wing's Ist Group (I./KG 77) and Major von Frankenberg was sent to the Eastern Front, where he was appointed group commander of the bomber wing KG 51.

KG 51 "Edelweiss" was, by then, involved in difficult operations during the Battle of Stalingrad, suffering heavy losses. The previous commander, Colonel Heinrich Conrady, had assumed his position on January 3, 1943; five days later, he was shot down and killed. To von Frankenberg, leading KG 51 was like a *déjà vu*: during his three first weeks at the unit, 21 aircraft were lost. KG 51 was operating from the Rostov airbase, but on February 5, 1943, it had to be evacuated in a panic as Soviet tanks surprisingly appeared near the airbase. That month, KG 51 lost forty percent of its aircraft, and all its group commanders.

On May 1, 1943, von Frankenberg landed his undamaged Ju 88 behind the Soviet lines. He had all the headquarter documents from KG 51 with him. As a reason for this landing in hostile territory, von Frankenberg himself would claim "a magnetic anomaly"—but already during the evening the next day, he was speaking on Radio Moscow, calling on the Germans to lay down their arms.

Von Frankenberg joined, as did von Einsiedel, *Nationalkomitee Freies Deutschland* and was later involved in forming the similar *Bund Deutscher Offiziere BDO*, the German Officers' Union, for anti-Nazi German officers in Soviet captivity. Just like von Einsiedel, von Frankenberg was kept in Soviet captivity after the end of the war, and when he was released in 1948, he was only allowed to settle in East Germany. But there, he settled well. He joined the NDPD, a sort of semi-liberal party that supported the governing SED party in East Germany, and there, he became a member of the board and reached a high position in the *Landtag* of Thuringia. After the fall of the Berlin wall, the NDPD merged with the former West German liberal party FDP. By then, the now 81-year-old von Frankenberg opted to retire. He died in Berlin on March 15, 2000.

Another group commander of a German bomber group who switched sides in 1943 was Colonel Walter Lehwess-Litzmann. He was not a nobleman as were the other two, but, as a matter of curiosity, it can be mentioned that his wife, Ingeborg, was the grandchild of Professor Lothar von Meyer, one of the designers of the periodic table. Lehwess-Litzmann had flown many missions over England and on the Eastern Front as commander of the bomber group III./KG 1 when he was appointed, in January, 1943—at the same time as von Frankenberg took over KG 51—commander of the bomber wing KG 3 "Blitz."

KG 3, too, suffered serious losses on the Eastern Front during 1943. During Lehwess-Litzmann's time as commander, KG 3 twice lost its entire stock of aircraft, as well as 204 killed or missing and 62 wounded airmen. In addition, the ground crew at the airbase had been infiltrated by partisans. On September 7, 1943, a bomb that had been smuggled aboard Lehwess-Litzmann's Junkers 88 exploded. The crew ended up in captivity. On October 29, 1943, Lehwess-Litzmann was awarded the Knight's Cross posthumously (he had been written off as missing). This, however, was something that Hitler would come to regret; for it did not take long before Lehwess-Litzmann reappeared—now as a member of the National Committee and the BDO!

For some reason, Lehwess-Litzmann was released soon after the end of the war, and settled in East Berlin, where he took part in founding the *Berliner Zeitung as early as in 1945*. A few years later, he joined the East German air force, and between 1959 and 1970, he was the head of the East German airline Interflug. Following a car accident where he was badly injured in 1970, he was given disability pension and died in 1986. In 1994, his son Jörn Lehwess-Litzmann published his father's autobiography, under the title *Absturz ins Leben* ("I Crashed into Life," not published in English).

In fact, quite a number of Luftwaffe airmen switched sides during the war. In most cases, they defected to the Soviets, after having (involuntarily) ended up in captivity. About 45 percent of the German officers and 75 percent of the privates in Soviet captivity allegedly joined the National Committee.[127] In some of these "change-over" cases, the question is whether or not it had been planned in advance.

On June 25, 1941, three days after Hitler's invasion of the Soviet Union, a Junkers 88 bomber from KG 54 belly-landed in a field north of Kiev. The German crew—the pilot, Lance Corporal Hans Hermann, the scout, Lance Corporal Hans Kratz, the radio operator, Lance Corporal Adolf Appel, and the gunner, Sergeant Wilhelm Schmid—stepped out and surrendered to Soviet farmers working in the field. They said that they were against the attack on the Soviet Union and that they had dropped their bombs in the river Dnieper and intended to hand themselves over to the Soviet authorities. They were immediately detained by the Soviet security service, to which they repeated the same story. A few days later, flyers were dropped over the German lines with a request to the German airmen: "Airmen and soldiers, brothers, follow our example. Abandon the murderer Hitler and come over to us, to Russia." (The Russian version of this flyer is depicted at http://relicfinder.info/forum/viewtopic.php?f=26&t=2830.) Whether this was planned or whether it was a conversion under the gallows has not been possible to establish.

On March 4, 1942, the pilot Lieutenant Herbert Baumgartner from 27—with the observer, Corporal Karl Erich Moser, the radio operator, Lance Corporal Kurt Lees, and the gunner, Sergeant Josef Wagner onboard—landed his Heinkel 111 bomber in the Soviet airfield at Leninsk in Ukraine. A few days later, flyers were dropped over Kirovograd, where KG 27 was stationed, according to which Baumgartner spoke directly to his comrades, Lieutenants Krenz and Lessmaier, and First Lieutenants Wagner, Lohmann, and Klein, with a request to join him.

But one case is certainly a matter of intentional desertion: a Luftwaffe airman actually flew and landed voluntarily in Great Britain three times during the war! His name was Herbert Schmitt (but he called himself Heinrich).

In 1941, he was a pilot with the night fighter squadron 2./NJG 2. In the evening of May 20, 1941, First Lieutenant Schmitt took off with a twin-engine Dornier 217 night fighter from Aalborg, Denmark, and landed, according to some kind of agreement, undisturbed at RAF Scrampton near Lincoln. There, he handed over a sealed parcel in greatest secrecy to a high-ranking person within the British High Command and was then allowed to take off and return to his own base. This flight has been confirmed.[128] But what the parcel contained, and which purpose Schmitt's mission had, is still shrouded in mystery today; the British documents on the matter still seem to be classified. It has been speculated that the secret parcel that Schmitt handed over had to do with Rudolf Hess' flight to Scotland. Schmitt himself says in an interview, "This was part of the shadow wars that went on in those days. I wasn't the only German pilot who landed in England under mutual agreement, and several British pilots landed in Germany, which leading people on our side were aware of."[129]

In fact, Schmitt had been to England once before during the war: in the evening of February 14, 1941, he made similar flight to RAF Debden.[130] But there would be a third flight to a British airbase, this time, a definitive one. In 1943, Schmitt was stationed at the night fighter squadron 10./NJG 3. At 1503 hours on May 9, he took off from

Aalborg together with the radio operator, Master Sergeant Paul Rosenberger, and the flight mechanic, Master Sergeant Erich Kantwill. The flew a Junkers 88 R-1 night fighter, equipped with the very latest type of German aircraft radar, FuG 202 Lichtenstein B/C. Aided by this, the German night fighter command had been able to inflict increasing losses on the British Bomber Command during the last few months. The average loss quota for the nocturnal British bomber flights had now almost reached five percent, which meant that British bombers fliers could not expect to survive more than, at most, 20 sorties.

Schmitt flew to the airfield at Kristiansand, Norway, where he landed after exactly one hour's flight to refuel. At 1650 hours, the night fighter crew takes off again. Their mission is to track down and attack a British Mosquito courier plane over the Skagerrak, which was flying regularly between Stockholm and Leuchars by this time in order to, amongst other things, deliver Swedish ball bearings. But Schmitt, who had not shot down any aircraft until then, instead sets course straight out across the North Sea.[131]

The 29-year-old Schmitt was a very experienced pilot. He had been flying as a volunteer in the German Condor Legion in the Spanish Civil War. There, his experience was used to try out various types of bomber tactics with Junkers 86s, Heinkel 111s, and Dorner 17s.[132] But it was as he witnessed the terrible bombing of Guernica that his doubts began. As he later came home again, he was more receptive for arguments from his father, who was a Social Democrat and underground opponent to Hitler. When the Gestapo then arrested Herbert/Heinrich Schmitt's Jewish girlfriend and threw her into a concentration camp, it was the last straw.[133] From that moment on, Herbert/Heinrich was a sworn opponent to Hitler. After the war, he recounted, "I saw the condition of things: the oppression Germany, the murdering on the battlefields... My chick, who was a Jewess, was thrown into a concentration camp where she was killed. We were wading in blood—that could have to be enough!"[134]

The fact that the radio operator onboard Schmitt's aircraft, Paul Rosenberger, was an opponent of Nazism is not difficult to understand either; he was a Jew himself, but had managed to escape the Nazi authorities' attention and enlisted in the German Wehrmacht.[135] He was not alone in doing this. According to very interesting research by the prized American historian Bryan Mark Rigg, as many as 150,000 men with Jewish heritage served in the German Wehrmacht during World War II.[136] Several of these rose to the rank of general, and some were awarded the Knight's Cross—including some of the most successful fighter pilots of the Luftwaffe, even some with more than 100 aerial victories. Volume 2 of *Luftwaffe Pilots* will deal closely with this subject.

On the other hand, Kantwill, the flight mechanic, knew nothing about what the two others were planning. At 1710 hours, Rosenberger broadcasts a false radio message to the night fighters' supreme command in Denmark that the starboard engine had caught fire. Kantwill protested wildly, but then calmed down after Rosenberger had threatened him with his service pistol. Schmitt now takes the Junkers plane down to right above the waves to escape the German radar, and then, Kantwill, threatened by Rosenberger's gun, throws out three rubber life rafts. Schmitt's and Rosenberger's intention is to make the other Germans believe that the aircraft had actually crashed over the North Sea. Then, Schmitt continues, heading for Scotland.

Less than two hours later, the aircraft is intercepted by the British radar station at Peterhead, more than thirty kilometers north of Aberdeen, and from the fighter squadron 165 Squadron in Dyce, two Spitfires take off, with the pilots Flight Lieutenant Arthur Roscoe and Sergeant Ben Scamen. 165 Squadron's war diary describes the mission, "Blue Group was ordered by Peterhead's sector command (Flight Lieutenant Crimp) to intercept an incoming aircraft. This had been located to east of Peterhead, but it turned southwards and flew for some time along the coast, eventually starting to circle a few *miles* north-northwest of Dyce. The fighters were led in towards this aircraft, which was identified as a Ju 88. The hostile aircraft deployed its landing gear, fired a few flares, and started rocking its wings violently as Flight Lieutenant Roscoe approached. He responded in the same way and positioned himself in front of the enemy aircraft in order to lead it to [the airbase] Dyce.

Blue 1 ordered Blue 2 to position himself behind and above the Junkers machine, and the whole company continued towards Dyce where they all landed. The pilots should be commended for not opening fire, but bringing valuable information to the technical department instead, and the fighter commander is to be commended for his quick understanding of the situation and his leadership during it."[137]

The Junkers 88 is surrounded by military police as it lands, and the crew calmly steps outside. They surrender to the commander of the airbase, Group Captain David Colquhon, who is surprised to discover that two of the Germans are wearing their handsome blue dress uniforms underneath their oil-stained flight coveralls.

According to Flight Lieutenant Charles Sharp, who by then was serving at RAF Dyce, the three Germans were allowed to stay at the airbase for a week, living at the officers' mess during this time.

A few days after the landing, the British agreed to send a coded radio message through the British secret radio station Gustav Siegfried Eins: "May has arrived (*Der Mai ist gekommen*)." According to Schmitt, this was the agreed confirmation for his father that he had landed safely in Great Britain. In the Luftwaffe, on the other hand, the aircraft was reported crashed into the North Sea.

After a while, Schmitt and Rosenberg started broadcasting radio calls under assumed names to Germany over Gustav Siegfried Eins at 1600 hours every day. "The war is lost," they said. "Do not sacrifice your lives for a pointless war with incompetent leaders. In England and Sweden, there are airfields where you will be treated as well as we have been. Remember—rock your wings, and you will be escorted to a safe landing."[138]

By coming across the advance German flight radar onboard the Junkers 88—and the extensive collection of documents about this equipment that Schmitt had taken with him—the British were able to develop several countermeasures that would cut their losses drastically. One of these was the Serrate radar detection and homing device that was fitted into British aircraft and which reacted to signals from Lichtenstein

B/C. As early as in June, 1943, the loss rate for RAF Bomber Command dropped from almost five percent to a little over a mere three percent. With the captured German radar equipment, the British were also able to measure the optimum length of the chaff (so-called *Windows*) that they would soon deploy in large quantities to "blind" the German radar. The British used *Windows for the first time in connection with* Operation "Gomorrah," the bomb attack on Hamburg in July 1943. The outcome was that the German radar was completely knocked out, and 42,600 people died in Hamburg during the most concentrated attack by the British bomber command until then.

Schmitt returned after the war to West Germany and was eventually hired as a test pilot by Triumph but emigrated after a few years after which all traces of him are gone. According to the German agency for information on former Wehrmacht soldiers (Deutsche Dienststelle WASt), "the pilot First Lieutenant Herbert Schmid" died on May 4, 1983.[139] Rosenberger, however, remained in Great Britain, where he assumed a new identity. In 1979, he was running a hotel and a restaurant in Malborough, Wiltshire. Erich Kantwill emigrated to Canada and then to the USA. But their Junkers 88 R-1 still remains in Great Britain and is nowadays on display at the air museum RAF Hendon in London.

Heinrich Graf von Einsiedel (l.) And his group commander, Wolfgang Ewald. (Photo: von Einsiedel.)

Soviet fighter ace Mikhail Baranov in front of his Yak-1 fighter aircraft in the summer of 1942. Each star represents a shot down enemy fighter aircraft. The text above the stars reads, freely translated: "The Fascists' Fear, M. D. Baranov." (Photo: Kovachevich)

*Walter Lehwess-Litzmann.
(Photo: von Einsiedel.)*

The crew of the Junker 88 bomber from KG 54, which on June 25, 1941 belly-landed behind the Soviet lines, claiming that they wanted to desert to the Soviet side. From left: Corporal Hans Kratz, Lance Corporal Wilhelm Schmid, Corporal Adolf Appel, and Corporal Hans Hermann. Was it a deliberate desertion or a "conversion under the gallows"? Clip from the Soviet newspaper Krasnaya zvezda ("The Red Star").

The wrecked of a crashed Junkers 88. (Photo: Mombeek.)

A Heinkel 111 of bomber wing KG 27 at Stalino airbase in the spring of 1942. (Photo: Broschwitz/ Traditionsgeschwader JG 52.)

Today, this Junkers 88 R-1 that Heinrich Schmitt and Erich Kantwill flew to the British on May 9, 1943 is on display at the RAF Museum Hendon in north London. (Photo: The author.)

The Danish volunteer fighter pilot in the Luftwaffe, Knud Erik Ravnskov. (Photo: Ravnskov.)

The gun camera on another aircraft has captured how an American P-47 Thunderbolt sets a Focke-Wulf 190 on fire. (Photo: NARA, 57363A.C.)

The Danish Luftwaffe pilot Captain Ove Terp in the cockpit of his Me 109 in JG 54 at the airfield Malmi in Finland in 1943. (Photo: Trautloft.)

CHAPTER 6

Danish Pilots in the Luftwaffe

The title of this book is *German Airmen During World War II*, but it could still be interesting to take a closer look at the Danish pilots who flew in the Luftwaffe.

When Pilot Officer John Newton Miller from the RAF's 185 Squadron broadcast a radio message in the afternoon of March 20, 1943, that his Spitfire had been shot in flames over the Mediterranean off Sicily, a rescue mission was launched. From its base in Malta, a Walrus-type amphibious plane took off to save the pilot out of the freezing waters. But since both sides from time to time used to shoot down each other's air-sea rescue planes (which were suspected of also carrying out military reconnaissance missions), the amphibious plane was given an escort of four Spitfire fighters.
At the same time, two Messerschmitt 109s took off from an airbase in Sicily. In order for the crew to have a chance to spot the distressed pilot, the Walrus plane could not fly at the lowest altitude but was forced to climb straight upwards to an altitude where the German radar would also pick them up. Therefore, the two Me 109s could be led straight towards the British planes, and they were also able to position themselves higher than their opponents. The Messerschmitt pilots spotted the British just as the amphibious plane had landed on the waters to pick up Miller.
 The Spitfire pilots did not discover the danger until it was too late. The Me 109s descended with the sun in their backs, shot down one

of the Spitfires, and then left. It was the kind of "shock attack" that the Luftwaffe's fighter pilots had specialized in. Thus, Captain Poul Sommer from the Luftwaffe's fighter wing JG 27 had been credited with his third aerial victory. The remarkable thing about Sommer was not only that he was unusually old for being a fighter pilot in the Luftwaffe, almost 33 years old, but he was also a Danish citizen.

By this time, Poul Sommer was one of the most experienced pilots at JG 27. Aged 24, he had joined the Danish navy's air force in 1934, where he trained to become a fighter pilot and was promoted to First Lieutenant. But when he reported for duty in the Soviet-Finnish Winter War, which broke out in November, 1939, the Danish naval staff refused to give its approval.

The conditions changed quickly with the German invasion of Denmark on April 9, 1940. The Danish government's policy was to mitigate the effects of the occupation through collaboration with Germany. Amongst other things, Denmark joined the Anti-Comintern Pact and accepted in 1941 that Danish citizens could volunteer to participate in the war on Germany's side. Among several Danish officers who volunteered was Poul Sommer. Most of them joined the SS, but to Sommer, it was natural to apply to the Luftwaffe.

Having had thorough training on the Messerschmitt 109, First Lieutenant Sommer was stationed in "Edu" Neumann's JG 27 in North Africa in September, 1942. Soon after his arrival, "Jochen" Marseille is killed in an air accident.

On October 23, 1942, General Montgomery's British Eighth Army launches its frontal attack at el-Alamein, while hundreds of Allied aircraft are pounding the Axis positions. Fighter pilots from JG 27 are incessantly in action to defend their badly beleaguered ground troops.

In the afternoon of this October 23, Sommer is participating as eighteen German and Italian fighters launch an attack on a large British aircraft en route to attack the Axis airfields. In a cauldron of turning and firing aircraft, the Danish first lieutenant manages to score a full series of hits into a British fighter, which he sees crashing to the ground.

It is quite possible that Sommer came across a fellow Dane during the bitter aerial fighting that was raging during the Battle of el-Alamein. One of the bomber units that took part on the Allied side was the RAF's No. 14 Squadron, and in this unit, the Danish pilot Gunnar Christensen Egebjerg served at the same time.

On October 28, 1942, Egebjerg and the other airmen at No. 14 Squadron became the first ones to deploy the new American-made Marauder bombers in combat—right at el-Alamein. At the exact same moment, Poul Sommer was one of ten Me 109 pilots deployed against a large formation of Allied bombers over el-Alamein. Thanks to the British fighter escort, the Me 109s failed to get through and attack the bombers, but Sommer shot down one of the escorting Spitfire planes. The pilot flying this, Flight Lieutenant Cocker, had survived a downing by parachuting on July 17, 1942, but this time, he would not make it.

The Battle of el-Alamein, however, was lost. Moreover, a new serious threat arose as Allied forces landed in Morocco and Algeria in November 1942. In order to meet this, units including II./JG 27, the fighter group that Sommer belonged to, were transferred to Sicily. It was while operating from there that Sommer and his comrades got to feel the growing Allied superiority in numbers in the air.

Sommer saw one of his comrades after another crashing in their burning aircraft. Eventually, Sommer himself gets injured in combat.

In September 1943, Sommer is back in Denmark. A few months later, he is designated commander of the Danish guards responsible for watching over German airbases in Denmark. His final result as a fighter pilot will remain three aerial victories.

With his activities for the occupying power, Sommer drew the attention of the Danish Resistance, and on August 15, 1944, he had a gunshot wound during an attack outside his home in Charlottenlund.

After the war, Sommer was subject to a legal trial of his services for the occupying power. In an initial trial, he was sentenced to eight years in prison, a penalty that the Supreme Court would later raise to twelve years' imprisonment. However, he was pardoned after a few years, retiring to a quiet life as a businessman.

Sommer was one of seven Danes who volunteered in the Luftwaffe. Six of these were admitted and two of them served in the same unit, JG 54 "Grünherz": Ejnar Thorup and Ove C. Terp. Thorup was born on December 7, 1923, in Skanderup, and trained as a fighter pilot in 1936. He was able to serve as a volunteer fighter pilot in Finland during the last few days of the Winter War. When Germany invaded Denmark on April 9, 1940, Thorup was stationed at Sjællandske Flyverafdeling (the Zealand Air Wing).

Since Denmark was allowed to maintain its sovereignty, Thorup was able to continue serving in the Danish air force. On May 22, 1940, he was appointed an adjutant to the chief inspector of the air force, Colonel Førslev. But when the Danish War Ministry announced in July, 1941, that it was permitted for Danish officers to volunteer with the Nazi Frikorps Denmark, Thorup hurried to volunteer with the Luftwaffe. In August, 1941, he was transferred to fighter pilot school Jagdfliegerschule 5 (where Hans-Joachim Marseille had had his training as a fighter pilot a little over a year earlier), and on May 29, 1942, he was sent, together with Ove Terp, to JG 54 "Grünherz" on the Eastern Front.[140]

With the rank of captain, Thorup was stationed at the 5th Squadron (5./JG 54), but he did not last long there. After only one week, he was shot down by the Soviets. Thorup attempted an emergency landing, but his aircraft crashed to the ground. The pilot and his seat were thrown out of the cockpit. When the rescue mission arrived on the scene, Thorup was dead.

On January 18, 1943, the commander of JG 54, Hannes Trautloft, received a letter from Ejnar Thorup's widow: "*Herr* Major H. Trautloft. Only now have I been able to compose myself enough to respond to both of your letters. Of course, it made me happy to learn that my husband also was a good comrade and soldier with you. I can only say that he was the best man in the world, and to me, life has no meaning anymore after his death. I only remain alive so that his children will not become alone in this world. I would actually like to hear more about the time when my husband was with you and speak to people he associated with. But it is of course impossible; but if you, *Herr* Major, ever come to Denmark, I would be grateful if you would like to visit me. Once again, I thank you for your two letters and greet you. Heil Hitler."[141]

The other Danish pilot at JG 54, Ove C. Terp, managed considerably better. Born in 1914 and trained as a pilot in Denmark, he volunteered in the Luftwaffe at the same time as Ejnar Thorup and was also sent to Jagdfliegerschule 5. He was posted to III./JG 54, but, strangely enough, there seems to be no documents on him. Trautloft remembered that the Danish Second Lieutenant Terp reported to his unit late in the spring of 1942, and that he soon afterwards had surgery for appendicitis, but he cannot remember what happened to Terp after that.[142] No other veterans from JG 54 whom I have spoken to seem to have any memory of him. Could it be that Terp's tenure at JG 54—perhaps also in the Luftwaffe—became as short as Thorup's? What we do know is that Terp emigrated to the USA after the war and was hired as a flight mechanic in the US Air Force. In 1950, he became a West German citizen, and served in the Bundesluftwaffe, where he retired a Lieutenant Colonel on March 26, 1979.

Two of the Danish pilots served with JG 51, which was given the honorary name "Mölders" after Werner Mölders' death in November, 1941. Wolfgang Fabian, born in 1915, was in the same pilots' class as Poul Sommer, and received his "wings" in 1936. In the summer of 1938, he became famous for his participation in the Danish Cartographic Institute's flying research expedition to the interior of Greenland. When Germany invades Denmark in 1940, Wolfgang Fabian is a first lieutenant at Luftmarinestation Slipshavn near Nyborg in Funen. One of his unit comrades was Kaj Birksted. In the evening of April 16, Birksted fled to Sweden by boat, and from Sweden, he made it to Great Britain. There, he was deployed in the RAF, flew a Spitfire, and became, with 10½ aerial victories, officially Denmark's most successful fighter pilot.

But Wolfgang Fabian went the other way and volunteered in the Luftwaffe in November 1941. After receiving his fighter training at Jagdfliegerschule 4, he was sent, with the rank of First Lieutenant, to the 9th Squadron of JG 51 on the Eastern Front. Not much is known about

Fabian's tenure at JG 51, and he is not registered for any aerial victories. On August 21, 1942 he was shot down in an Me 109 F-2 northeast of Rzhev and was killed.

It is possible that Peter Horn, the other pilot at JG 51, who was the same age as Fabian, surpassed Kaj Birksted's results as a fighter pilot. He, too, was a flight officer, first lieutenant, in the Danish air force when Germany invaded and he, too, volunteered in the Luftwaffe as early as during the summer of 1941. There is not much information on Horn either, but he allegedly served with the fighter squadron 1./JG 51 from late 1941 and was credited with ten or eleven aerial victories. Peter Horn survived the war and died on November 1, 1983.

On the other hand, there is much more information on the Danish Luftwaffe airman Knud Erik Ravnskov, and that is thanks to the thorough research that his grandchild Pernille Ravnskov has done. Knud Erik Ravnskov, born in 1911, was a captain lieutenant at Jydske Flyverafdeling when the Germans came marching in. In August, 1941, he took the train to Berlin and reported for duty at the Luftwaffe. Only a few days later, he too was sent to Jagdfliegerschule 5.

But Ravnskov suffered badly from rheumatism and had to be admitted to hospital in Berlin. On November 23, 1941, he was sent back to Jagdfliegerschule 5, but was transferred in January 1942 to Jagdfliegerschule 1. The idea was that Ravnskov, too, would be sent to JG 51 "Mölders." In the spring of 1942, he was stationed at the 1st Squadron in the replacement group Ergänzungsjagdgruppe Ost, which provided JG 51 with new pilots. But once again, he had to be admitted to hospital because of his rheumatism.

Only in late August 1942 was Ravnskov discharged from the hospital. On September 14, 1942, he reassumed flight training with Me 109s. Three days later, Knud Erik Ravnskov crashed in an Me 109 on the Krakow airfield and was badly injured. He died the next day.[143]

The oldest one among the Danish Luftwaffe pilots was Anker Tage Harild (born in Copenhagen in 1905). He was also a captain lieutenant at the time of the German occupation. Harild was a member of the Danish Nazi Party and joined the Luftwaffe in 1941, where he was considered to be most suitable as a bomber. He flew Junkers 88 bombers with the bomber group III./KG 1 under Major Walter Lehwess-Litzmann's command. Not much is known about him either, but he was allegedly still at the unit in 1943 and reached the rank of major. Anker Tage Harild survived the war and was suspended from the Danish armed forces on May 25, 1945 and was sentenced to six years in prison for his collaboration with the Germans.

However, his fate was better than the one for the sixth and final one of these Danish airmen in the Luftwaffe, Uffe Børge Darr. He was actually Danish-German. Darr was born in Silkeborg, Denmark in 1919, but moved with his parents before his first birthday to Germany, where he eventually became a German citizen. As an adult, he married a Dane, however, and then regained his Danish citizenship.

Through his previous German citizenship, Børge Darr was able to be admitted to the Luftwaffe as early as in 1940. After having gone through the A, B, and C flight schools, he was placed as a flight instructor at a pilot training school. On November 11, 1943, he had his first deployment to a combat unit, the fighter group II./JG 11 in the Home Defense. Other notable pilots at this unit included Heinz Knoke, well-known for his memoirs.

It is unknown whether or not Darr took part in the five attacks that II./JG 11 launched on American heavy bombers during the rest of November—whereby the unit was credited with the downing of three bombers and three fighters against four own losses—but when U.S. Eighth Air Force deployed 299 four-engine bombers on December 1, 1943, with an escort of more than 400 fighters against industries in the Ruhr area, Darr was one of the around three hundred German fighter pilots taking off to fight back.

While the more heavily armed Focke-Wulf 190s were focusing on fighting the bombers, the Me 109-equipped II./JG 11 was tasked with protecting these from the fighters. I. and III./JG 1 carried out a first frontal attack against a formation of thirty-three B-24 Liberator bombers southwest of Bonn, and shot down two. But after that, the American escort fighters descended upon the Germans, and II./JG 11 engaged.

The American P-47 Thunderbolt pilot Lieutenant James "Wilkie" Wilkinson from the 78th Fighter Group describes what was conclusively Børge Darr's last aerial fight, "Yellow Flight, of which I was #4, was attacked by four 109G's at 28,000 ft. Yellow leader waited until the attack developed and broke sharp left, the 109's following. Three of the enemy aircraft singled me out for their attack but two of the three broke off as I continued my left turn. The third enemy aircraft continued to press the attack down to about 20,000 ft where he was either driven off or chose to break off the attack. Seeing my tail cleared, I started to climb back to rejoin my flight when I was attacked by the same Hun. On being attacked, I broke down and to the right and continued to turn right down to about 8,500 ft at which point I was beginning to turn inside the Hun, so he tried to break off the attack by diving away to the right. I followed his maneuver and, thinking that I would not be able to close on him, I took a deflection shot at extreme range. The Hun apparently thought I could not or would not close with him—as he allowed me to get dead astern, from which position I took about three short shots at him, until it became apparent that I was closing on him quite rapidly (before he could reach cloud cover below) so I held my fire until the range closed to about 100-150 yards. At this point, he appeared to explode in a large burst of white smoke. I observed numerous strikes on the enemy aircraft from the range of about 450 yards until he finally exploded. During the first phase (20,000 down to 8,500) he fired several times but did not succeed in getting any strikes on me. This combat took place at approximately 1125 hrs and lasted eleven or twelve minutes."[144]

Returning to base, it was concluded that II./JG 11 had shot down two P-47 Thunderbolts. Børge Darr was the unit's only loss. (78th Fighter Group also lost two Thunderbolts in that battle.) But this aerial combat still reflects the beginning of the end of the Luftwaffe. With the Republic P-47 Thunderbolt, the Allies had an aircraft which was manifestly superior to the most common German fighters. At an altitude of 7,000 meters, roughly the flight level where most of the aerial battles played out over Germany during the fall/winter of 1943 and the decisive year 1944, the P-47D had a top speed of 683 km/h, while the Me 109 G-6 could reach 645 km/h and the Fw 190 A-8 635 km/h. In addition, the Thunderbolt was able to dive away easily from both the German machines, which meant that the most common German evasive maneuver for fighters in aerial combat—an *Abschwung* (see p. 19)—was not of much use.

The same month as Børge Darr was killed, another American fighter came into service in Europe, the North American P-51 Mustang. This was even more superior to the Germans' single-engine fighter, not only through its outstanding range—normally 1,500 kilometers, 3,700 kilometers with extra tanks—but by virtually all parameters: at an altitude of 7,000 meters, the P-51B had a top speed of 716 km/h; it was thus 80 km/h faster than the Fw 190—which could be compared with the fact that the German Me 109 F-4 in 1941 was 95 km/h faster than the old Soviet Polikarpov I-16 fighters (which the Germans called "Rata"). Neither did the Mustang have any difficulties outmaneuvering any of the two German fighters, it was able to easily dive away from them both, and climbed faster than they did, at least above 6,000 meters.

With the P-51 Mustang—which, during 1944, became the American standard fighter in aerial combat over Germany—the Luftwaffe's fighters finally lost the battle for air supremacy. With that, the mass death of German pilots began.

APPENDIX 1

German Fighter Pilots with 100 or more Aerial Victories during World War II

Name	Number of aerial victories	Name	Number of aerial victories
Erich Hartmann	352	Johannes Steinhoff	176
Gerhard Barkhorn	301	Ernst-Wilhelm Reinert	174
Günther Rall	275	Günther Schack	174
Otto Kittel †	267	Emil Lang †	173
Walter Nowotny †	258	Heinz Schmidt †	173
Willi Batz	237	Horst Ademeit †	166
Erich Rudorffer	222	Wolf-Dietrich Wilcke†	162
Heinz Bär	220	Hans-Joachim Marseille †	158
Hermann Graf	212	Heinrich Sturm †	158
Heinrich Ehrler †	208	Gerhard Thyben	157
Theodor Weissenberger	208	Hans Beisswenger †	152
Hans Philipp †	206	Peter Düttmann	152
Walter Schuck	206	Gordon Gollob	150
Anton Hafner †	204	Fritz Tegtmeier	146
Helmut Lipfert	203	Albin Wolf †	144
Walter Krupinski	197	Kurt Tanzer	143
Anton Hackl	192	Friedrich-Karl Müller †	140
Joachim Brendel	189	Karl Gratz	138
Max Stotz †	189	Heinrich Setz †	138
Joachim Kirschner †	188	Rudolf Trenkel	138
Kurt Brändle †	180	Franz Schall †	137
Günther Josten	178	Walter Wolfrum	137

Name	Number of aerial victories
Horst-Günther von Fassong †	136
Otto Fönnekold †	136
Karl-Heinz Weber †	136
Joachim Müncheberg †	135
Hans Waldmann †	134
Johannes Wiese	133
Alfred Grislawski	132
Adolf Borchers	132
Adolf Dickfeld	132
Erwin Clausen †	132
Wilhelm Lemke †	131
Gerhard Hoffmann †	130
Heinrich Sterr †	130
Franz Eisenach	129
Walther Dahl	128
Franz Dörr	128
Rudolf Rademacher	126
Josef Zwernemann †	126
Dietrich Hrabak	125
Wolf Ettel †	124
Herbert Ihlefeld	123
Wolfgang Tonne †	122
Heinz Marquardt	121
Heinz-Wolfgang Schnaufer	121
Robert Weiss †	121
Friedrich Obleser	120
Friedrich Wachowiak †	120
Erich Leie †	118
Franz-Josef Beerenbrock	117
Hans-Joachim Birkner †	117
Jakob Norz †	117
Walter Oesau †	117
Heinz Wernicke †	117

Name	Number of aerial victories
August Lambert †	116
Wilhelm Crinius	114
Werner Schroer	114
Hans Dammers †	113
Berthold Korts †	113
Kurt Bühligen	112
Helmut Lent †	110
Kurt Ubben †	110
Franz Woidich	110
Reinhard Seiler	109
Emil Bitsch †	108
Hans Hahn	108
Bernhard Vechtel	108
Viktor Bauer	106
Werner Lucas	106
Günther Lützow	105
Adolf Galland	104
Eberhard von Boremski	104
Heinz Sachsenberg	104
Hartmann Grasser	103
Siegfried Freytag	102
Friedrich Geisshardt	102
Egon Mayer †	102
Max-Hellmuth Ostermann	102
Josef Wurmheller	102
Herbert Rollwage	102
Werner Mölders	101
Rudolf Miethig †	101
Josef Priller	101
Ulrich Wernitz	101

† = Killed in action.

APPENDIX 2

The True Results of the German Fighter Aviation

The table below does not display reported aerial victories (claims) but is based on a careful study of Allied aircraft losses in aerial combat, and constitutes an estimate of the true results of the German Fighter Aviation—beyond exaggerated claims.

Year	In combat with the Western Allies		On the Eastern Front	
	Enemy aircraft shot down	German fighters shot down	Enemy aircraft shot down	German fighters shot down
1939-1940	1,500	800		
1941	1,500	300	5,000	600
1942	2,500	500	9,000	600
1943	3,000	2,000	9,000	800
1944	5,000	8,000	7,000	1,100
1945	500<	1,000	2,000	1,500
Totals	14,000	12,500	32,000	4,500

APPENDIX 3

The Most Successful German Fighter Groups

The most successful German Fighter Wings were those operating on the Eastern Front during most of the war: JG 51 "Mölders," JG 52, and JG 54 "Grünherz." The most successful among these was JG 52, with a total of about 11,000 aerial victories.

The development of the successes for these fighter groups was as follows:

JG 52

Achieved aerial victory	Date
No. 500	Sept. 7, 1941
No. 1,000	Feb. 1942
No. 2,000	June 22, 1942
No. 4,000	Dec. 7, 1942
No. 5,000	April 20, 1943
No. 6,000	July 7, 1943
No. 7,000	Sept. 16, 1943
No. 8,000	Dec. 4, 1943
No. 9,000	May 10, 1944
No. 10,000	Sept. 2, 1944

JG 54

Achieved aerial victory	Date
No. 1,000	Aug. 1, 1941
No. 2,000	April 4, 1942
No. 3,000	Sept. 14, 1942
No. 4,000	Feb. 19, 1943
No. 5,000	July 17, 1943
No. 6,000	Oct. 9, 1943
No. 7,000	March 23, 1944
No. 9,000	Oct. 15, 1944

JG 51

Achieved aerial victory	Date
No. 2,000	Sept. 8, 1941
No. 3,000	April 7, 1943
No. 4,000	Nov. 2, 1942
No. 5,000	June 2, 1943
No. 6,000	July 27, 1943
No. 7,000	Sept. 16, 1943
No. 8,000	May 4, 1943

APPENDIX 4

The Most Successful Fighter Pilots Against Various Types of Aircraft

The aircraft that seem to have been the most demanding to shoot down were, judging by most Luftwaffe veterans' accounts (even though there are exceptions), the four-engine American B-17 Flying Fortress and B-24 Liberator bombers, which were usually armed with 12 heavy machine guns each, and which also flew in large, tight formations. Twenty-five German fighter pilots were credited with shooting down 20 or more four-engine bombers.

Name	Total number of aerial victories	Of which were against four-engine bombers
Georg-Peter Eder	78	36
Anton Hackl	192	34
Konrad Bauer	57	32
Walter Dahl	128	30
Werner Schroer	114	26
Egon Mayer	102	26
Rolf Hermichen	64	26
Hermann Staiger	63	25
Anton-Rudolf Piffer	35	26
Hugo Frey	32	25
Alwin Doppler	29	25
Kurt Bühligen	112	24
Hans Ehlers	55	24

Cont'd overleaf

Cont'd.

Name	Total number of aerial victories	Of which were against four-engine bombers
Friedrich-Karl Müller	140	23
Heinrich Wurzer	26	23
Walter Loos	38	22
Hans Weik	36	22
Werner Gerth	27	22
Heinz Bär	221	21
Fritz Karch	47	21
Willi Unger	24	21
Josef Wurmheller	102	20<
Willi Kientsch	53	20
Hans-Heinrich Koenig	28	20
Willi Reschke	27	20

Another aircraft which was considered difficult to shoot down was the heavily armored Soviet assault aircraft Ilyushin Il-2 Shturmovik. Eight German fighter pilots were still credited with shooting down 50 or more Il-2s. It is worth noting that all of them, with the exception of Johannes Wiese, flew Focke-Wulf 190s, which were more heavily armed than the Me 109.

Name	Total number of aerial victories	Of which were against Il-2
Otto Kittel	267	94
Joachim Brendel	189	88
Johannes Wiese	133	70<
Franz Schall	133	61
Günther Josten	178	60
Erich Rudorffer	224	58
Anton Hafner	204	55
Franz Eisenach	129	52

The "favorite" among the opponents' aircraft among the Luftwaffe's veteran fighter pilots seems to have been the British Spitfire. There is no other fighter on the Allied side that they express such deep respect for. Five German fighter pilots were credited with shooting down 50 or more Spitfires.

Name	Total number of aerial victories	Of which were against Spitfires
Josef Priller	101	68
Josef Wurmheller	102	56+
Hans "Assi" Hahn	108	53
Egon Mayer	102	51
Adolf Galland	104	50

One of the most difficult opponents was the American P-51 Mustang fighter, which was considerably superior to most German fighters. Five German fighter pilots were credited with shooting down 10 or more Mustangs.

Name	Total number of aerial victories	Of which were against Mustang
Wilhelm Steinmann	44	12
Heinrich Bartels	99	11
Heinz Bär	221	10
Franz Schall	133	10
Wilhelm Hofmann	44	10

Source: Aces of the Luftwaffe by Peter Kacha
luftwaffe.cz

APPENDIX 5

Data for Some of the Most Important Fighter Aircraft During World War II

Data	Focke Wulf 190 A-8	Hawker Hurricane Mk IIC	Yakovlev Yak-9U	Lavochkin La-5FN	Messerschmitt Me Bf 109 E-4*	Messerschmitt Me Bf 109 G-6
Country	German	British	Soviet	Soviet	German	German
Commissioned	1941 (this version in 1943)	1937 (Mk I)	1942 (this version in 1944)	1942 (this version in 1943)	1937 (this version in 1940)	1937 (this version in 1942)
Length	8.84 m	9.84 m	8.60 m	8.67 m	8.64 m	8.5 m
Wingspan	10.49 m	12.19 m	9.74 m	9.80 m	9.87 m	9.93 m
Weight (loaded)	4,900 kg	3,950 kg	3,204 kg	3,265 kg	2,665 kg	3,148 kg
Engine power	1,700 hp	1,185 hp	1,500 hp	1,850 hp	1,100 hp	1,475 hp
Top speed	653 km/h	547 km/h	672 km/h	648 km/h	560 km/h	640 km/h
Range	900 km	965 km	675 km	765 km	765 km	850 km
Armament	4 20mm automatic cannons, 2 13mm machine guns	4 20mm automatic cannons	1 20mm automatic cannon, 2 12.7mm machine guns	2 20mm automatic cannons	2 20mm automatic cannons, 2 7.92mm machine guns	1-3 20mm automatic cannons, 2 13mm machine guns

* The abbreviation Bf denotes Bayerische Flugzeugwerke, which was the original maker of this aircraft, before the company was bought by Messerschmitt.

Data	Messerschmitt Me 262 A-1a**	MiG-3 (Mikoyan-Gurevich 3)	North American P-51 D Mustang	Polikarpov I-16 Isjak (typ 28)	Republic P-47 Thunderbolt	Vickers Supermarine Spitfire Mk IX
Country	German	Soviet	American	Soviet	American	British
Commissioned	1944	1941	1942	1934 (type 1)	1942	1939 (this version in 1942)
Length	10.60 m	8.25 m	9.83 m	6.13 m	11.00 m	9.54 m
Wingspan	12.50 m	10.20 m	11.28 m	9.00 m	12.42 m	11.23 m
Weight (loaded)	7,045 kg	3,355 kg	4,175 kg	1,941 kg	5,774 kg	4,310 kg
Engine power	8,8 kilonewton	1,350 hp	1,720 hp	1,100 hp	2,600 hp	1,660 hp
Top speed	870 km/h	640 km/h	703 km/h	525 km/h	713 km/h	657 km/h
Range	850 km	820 km	2,755 km with extra tanks	700 km	1,290 km	700 km
Armament	4 30mm automatic cannons, 24 rocket missiles	1 12.7mm machine gun, 2 7.62mm machine guns, 8 rocket missiles	6 12,7 mm machine guns	2 20 mm automatic cannons, 2 7,62 mm machine guns, 8 rocket missiles	8 12,7 mm machine guns	2 20 mm automatic cannons, 2 12,7 mm machine guns

** Messerschmitt Me 262 was the world's first operative jet fighter.

Sources
Archiv JG 54—Günther Rosipal.
National Archive, Kew.
National Archives and Records Administration, Washington, D.C.
RAF Museum Hendon.
Traditionsgemeinschaft JG 52, Engen.
TsAMO, Podolsk.
WASt Deutsche Dienststelle, Berlin.

German claims are based, unless indicated, on Abschusslisten, Film C2025N. 2026N, 2031N-2036N, Bundesarchiv/Militärarchiv, Freiburg.

Details on German aircraft losses are based, unless indicated, on Luftwaffe Aircraft Loss List by Matti Salonen.

Details on British aircraft losses are based, unless indicated, on their respective units ORB in National Archives, Kew.

Other unpublished material from private archives etc.
Various battle reports (ref. footnotes)
Various pilots' logbooks (ref. footnotes)
Trautloft, Hannes, diary.

Bibliography
Aders, Gebhard & Werner Held. *Jagdgeschwader 51 "Mölders": Eine Chronik.* Motorbuch Verlag, Stuttgart 1993.
Alexander, Kirsten. *Clive Caldwell Air Ace.* Allen & Unwin, Criws Nest NSW 2006.
Balke, Ulf. *Der Luftkrieg in Europa 1939-1941,* Bechtermünz Verlag, Augsburg 1997.
Barbas, Bernd. *Die Geschichte der II. Gruppe des Jagdgeschwaders 52,* Traditionsgemeinschaft JG 52, Überlingen 2005.
Bergström, Christer. *Graf & Grislawski: A Pair of Aces.* Eagle Editions, Hamilton MT 2003.
Bergström, Christer. *Black Cross/Red Star: Air War Over the Eastern Front, Volume 3: Everything for Stalingrad.* Eagle Editions, Hamilton MT 2006.
Bergström, Christer. *The Ardennes 1944-1945: Hitler's Winter Offensive.* Vaktel Books & Casemate, 2014.
Bergström, Christer. *The Battle of Britain: An Epic Battle Revisited.* Vaktel Books, Eskilstuna 2015.
Bergström, Christer. *Berömda flygaress och deras plan.* Vaktel förlag, Eskilstuna 2015.
Bergström, Christer. *Operation Barbarossa: Hitler Against Stalin 1941.* Vaktel Books & Casemate, 2016.

Braatz, Kurt. *Werner Mölders: Die Biographie*. NeunundzwanzigSechs Verlag, Moosburg 2008.

Brütting, Georg. *Das waren die deutschen Kampfflieger-Asse 1939-1945*. Motorbuch Verlag, Stuttgart 1975.

Bykov, Michail Ju. *Asi Velikoy Otechestvennoy: Samye resultativniye lyochiki 1941-1945 gg*. Jauza/Eksmo, Moscow 2007.

Caldwell, Donald. *The JG 26 War Diary, Volume One 1939-1942*. Grub Street, London 1996.

Caldwell, Donald. *The JG 26 War Diary, Volume Two 1943-1945*. Grub Street, London 1998.

Chapman Mede, Patricia. *The True Story of Catch 22: The Real Men and Missions of Joseph Heller's 340th Bomb Group in World War II*. Casemate, Philadelphia & Oxford 2012.

Clostermann, Pierre. *Det stora uppdraget*. Gleerups, Malmö 1961.

Cornwell, Peter D. *The Battle of France Then and Now: Six Nations Locked in Aerial Combat September 1939 to June 1940*. Battle of Britain International Ltd., Old Harlow 2007.

Crandall, Jerry. Illustrated by Tom Tullis. *Major Hans "Assi" Hahn: The Man and His Machines*. Eagle Editions, Hamilton MT 2002.

Düttmann, Peter "Bonifazius." *Wir Kämpften in einsamen Höhen*. Traditionsgemeinschaft Jagdgeschwader 52 & Falk Klinnert, Waizendorf 2002.

Graf von Einsiedel, Heinrich. *Tagebuch der Versuchung. 1942-1950*. Ullstein Zeitgeschichte, Mainz 1985.

Ewald, Heinz "Esau," *Wo wir sind ist immer Oben: Als Jagdflieger im Jagdgeschwader 52*. Traditionsgemeinschaft Jagdgeschwader 52 & Markus Ewald & Falk Klinnert, Waizendorf 1998.

Fast, Niko, *Das Jagdgeschwader 52*. Bensberger Buch-Verlag, Bergisch Gladbach 1988-1992.

von Forell, Fritz. *Mölders und seine Männer*. Steirische Verlagsanstalt, Graz 1941.

Foreman, John & Winfried Bock. *Air-War 1941: The Non-Stop Offensive, Part One*. Air Research Publications, Walton-on-Thames 2017.

Forsyth, Robert. *Jagdverband 44: Squadron of Experten*. Osprey Publishing, Oxford 2008.

Galland, *Die Ersten und die Letzten: Die Jagdflieger im Zweiten Weltkrieg*. Schneekluth Verlag, Munich 1970.

Gibbes, Bobby. *You Live but Once*. Collaroy, New South Wales 1994.

Gisclon, Jean. *Chasseurs au groupe "Lafayette" 1916-1945*. Nouvelles Editions Latines, Paris 1994.

Grunewald, Michel (publ.) & Uwe Puschner. *Das katholische Intellektuellenmilieu in Deutschland, seine Presse und seine Netzwerke (1871-1963)*. Peter Lang AG, Internationaler Verlag der Wissenschaften, Bern 2006.

Hagena, Hermann. *Jagdflieger Werner Mölders: Die Würden des Menschen reicht über den Tod hinaus*. Helios Verlag, Aachen 2008.

Halbig, Fabian. *Musikmissbrauch Im Dritten Reich: Musik als NS-Propaganda*. Grin Verlag, Munich 2016.

Hans, Hahn. *Ich spreche die Wahrheit!* Bechtle Verlag, Esslingen am Neckar 1951.

Hammel, Eric. *Air War Europa: America's Air War against Germany and North Africa 1942-1945*. Pacifica Press, Pacifica Ca. 1994.

Heaton, Colin D. & Anne-Marie Lewis. *The Star of Africa: The Story of Hans Marseille, the Rogue Luftwaffe Ace Who Dominated the WWII Skies*. Zenith Press, London 2012.

Helms, Bodo. *Von Anfang an dabei: Mein abenteuerliches Fliegerleben 1939-1980*. Kurt Vowinckel Verlag, Berg am Starnberger See n.d.

Hill, Robert. *The Great Coup: "Window," the Last Secret of World War II*. Corgi Books, London 1978.

Holmes, Tony. *American Eagles: US Fighter Pilots in the RAF 1939-1945*. Pen & Sword, Barnsley 2015.

Jones, Ira, *Tigrarna,* Hörsta förlag, Stockholm 1955.

Kondrat, Yemelyan Filaretovich. *Dostalsha namvek nespokojny,* DOSAAF, Moscow 1978.

Kurowski, *Hans-Joachim Marseille: Der erfolgreichste Jagdflieger des Afrikafeldzuges*. Flechsig-Buchvertrieb, Würzburg 2005.

Lehwess-Litzmann, Walter. *Absturz ins Leben*. Jörn Lehwess-Litzmann/Dingsda-Verlag, Querfurt 1994.

Longerich, Peter. *"Davon haben wir nichts gewusst!" Die Deutschen und die Judenverfolgung 1933–1945*. Siedler Verlag, Munich 2006.

Marks, Stephan. *Warum folgten sie Hitler? Die Psychologie des Nationalsozialismus*. Patmos Verlag, Düsseldorf 2007.

Mikoyan, Stepan Anastasovich. *Mikoyan: Memoirs of Military Test-Flying and Life with the Kremlin's Elite*. Airlife, Ilkey 1999.

Mobeeck, Erik & Jean-Louis Roba with Chris Goss. *In the Skies of France: A Chronicle of JG 2 "Richthofen," Volume 1: 1934-1940*. A.S.B.L. La Porte d'Hoves, Linkebeek n.d.

Morley-Mower, Wing Commander Geoffrey. *Messerschmitt Roulette: The Western Desert 1941-42*. Phalanx Publishing Co., St Paul, MN. 1993.

Neitzel, Sönke & Harald Welzer. *Soldaten: On Fighting, Killing and Dying—The Secret World War II Transcripts of German POWs*. Alfred A. Knopf, New York 2012.

Neulen, Hans Werner. *Am Himmel Europas: Luftstreitkräfte an deutscher Seite 1939-1945*. Universitas Verlag, Munich 1998.

Obermaier, Ernst. *Die Ritterkreuzträger der Luftwaffe 1939-1945. Band II, Stuka- und Schlachtflieger*. Verlag Dieter Hoffmann, Mainz 1976.

Obermaier, Ernst. *Die Ritterkreuzträger der Luftwaffe 1939-1945. Band I, Jagdflieger*. Verlag Dieter Hoffmann, Mainz 1989.

Paus, P. *Die Hölle von Hamburg: 1943—Vernichtung der Elb-Metropole durch alliierte Fliegerbombern.* Erich Pabel Verlag, Rastatt 1994.

Petermann, Viktor. *Der Jagdflieger Viktor Petermann: Pilot im Jagdgeschwader 52.* Traditionsgeschwader Jagdgeschwader 52 & OV Verlag/Falk Klinnert 2004.

Prien, Jochen, Gerhard Stemmer, Peter Rodeike & Winfried Bock. *Die Jagdfliegerverbände der Deutschen Luftwaffe 1934 bis 1945, Teil 4—Einsatz am Kanal und über England 26.6.1940 bis 21.6.1941*, Struve Druck, Eutin 2002.

Prien, Jochen, Gerhard Stemmer, Peter Rodeike & Winfried Bock. *Die Jagdfliegerverbände der Deutschen Luftwaffe 1934 bis 1945, Teil 5—Einsatz im Mittelmeerraum October 1940 bis November 1941, Einsatz im Westen 22. June bis 31. Dezember 1941.* Struve Druck, Eutin 2003.

Prien, Jochen, Gerhard Stemmer, Peter Rodeike & Winfried Bock. *Die Jagdfliegerverbände der Deutschen Luftwaffe 1934 bis 1945, Teil 7—Einsatz im Westen 1. Januar bis 31. Dezember 1942.* Struve Druck, Eutin n.d.

Prien, Jochen, Gerhard Stemmer, Peter Rodeike & Winfried Bock. *Die Jagdfliegerverbände der Deutschen Luftwaffe 1934 bis 1945, Teil 8—Einsatz im Mittelmeerraum November 1941 bis Dezember 1942.* Struve Druck, Eutin 2004.

Rigg, Bryan Mark, *Hitlers judiska soldater: soldater och generaler med judisk bakgrund i den nazistiska krigsmakten 1933-1945.* Vaktel förlag, Eskilstuna 2017.

Ring, Hans & Werner Girbig, *Jagdgeschwader 27: Die Dokumentation über den Einsatz an allen Fronten 1939-1945.* Motorbuch Verlag, Stuttgart 1972.

Ring, Hans & Christopher Shores. *Luftkampf zwischen Sand und Sonne: Luftkampf über Afrika 1940-1942.* Motorbuch Verlag, Stuttgart 1974.

Rybin, Jurij, *Boris Safonov: Luftwaffes baneman på östfronten.* Vaktel Books, Eskilstuna 2018.

Saunders, Andy. *Arrival of Eagles: Luftwaffe Landings in Britain 1939-1945.* Grub Street, London 2014.

Shores, Christopher & Clive Williams. *Aces High.* Grub Street, London 1994.

Shores, Christopher & Giovanni Massimello med Russell Guest. *A History of the Mediterranean Air War Volume One: North Africa June 1940-January 1942.* Grub Street, London 2012.

Shores, Christopher & Giovanni Massimello with Russell Guest, Frank Olynyk & Winfried Bock. *A History of the Mediterranean Air War 1940-1945. Volume Two: North Africa February 1942—March 1943.* Grub Street, London 2012.

Simonov, A.A. & N.G. Bodrichin. *Boyevyye lyochiki—dvazhdy i trizhdy Geroy Sovetskogo Soyuza.* Muzei Techniki Vadima Zadorozhnogo, Moscow 2017.

Sims, Edward H. *Jagdflieger—die grossen Gegner von Einst 1939-1945: Luftwaffe, RAF und USAAF im kritischen Vergleich.* Motorbuch Verlag, Stuttgart 1980.

Strauch, Dietmar. *Swing-Jugend und Edelweißpiraten: Biographien zum Widerstand im Dritten Reich.* Amazon Digital Services LLC 2017.

Tate, Robert. *Hans-Joachim Marseille: An Illustrated Tribute to the Luftwaffe's "Star of Africa."* Schiffer Publishing, 2008.

Theorell, Töres. *Noter om musik and hälsa.* Karolinska Institutet University Press, Stockholm 2009.

Thimmig, Max, *Nattens jägare: Tysk nattjaktflygare i andra världskriget.* Vaktel förlag, Eskilstuna 2015.

Waiss, Walter. *Boelcke-Archiv, Band III: Chronik Kampfgeschwader Nr. 27 Boelcke: Teil 2: 01.01.1941—31.12.1941.* Walter Waiss, Neuss, n.d.

Waiss, Walter. *Aus dem Boelcke-Archiv, Band IV: Chronik Kampfgeschwader Nr. 27 Boelcke: Teil 3: 01.01.1942—31.12.1942.* Helios Verlag, Aachen 2005.

Periodicals

52er Nachrichtenblatt.
Der Adler.
Aeroplane.
Bild am Sonntag.
Aeroplane Monthly.
Airlife.
AvStop Magazine Online.
Jägerblatt.
Kleine Kriegshefte. (1940)
Krasnaja zvezda.
The Mail on Sunday.
The New Zealand Herald.
Scramble! Official Newsletter of the Battle of Britain Historical Society.
South African Military History Journal.

Notes

1. Shores & Williams, *Aces High*, p. 358.
2. See e.g. Rybin, *Boris Safonov: Luftwaffes baneman på östfronten*, p. 57.
3. Interview with Alfred Grislawski.
4. Clostermann, *Det stora uppdraget*, p. 139.
5. Interview with Ilse Grislawski.
6. Shores & Williams, p. 358.
7. Günther Rall, logbook. Via Rall.
8. Johannes Steinhoff, logbook. Via Steinhoff.
9. Interview with Günther Rall.
10. Interview with Edmund Rossmann.
11. Interview with Walter Wolfrum.
12. Interview with Alfred Grislawski.
13. Johannes Steinhoff, logbook. Via Steinhoff.
14. Ring & Girbig, *Jagdgeschwader 27*, p. 238f.
15. Alfred Grislawski, logbook. Via Grislawski.
16. Interview with Klaus Häberlen.
17. Alfred Grislawski, logbook. Via Grislawski.
18. Interview with Hans-Ekkehard Bob.
19. Galland, *Die Ersten und die Letzten*, p. 98.
20. For a more detailed description of Adolf Galland, see the book *Berömda flygaress* by Christer Bergström.
21. Interview with Walter Wolfrum.
22. Ewald, *Wo wir sind ist immer Oben*, p. 15.
23. Interview with Wilhelm Batz.
24. Via Friedrich Lang.
25. Interview with Erhard Jähnert.
26. Interview with Alfred Grislawski.
27. Interview with Helmut Berendes.
28. Interview with Hugo Broch.
29. Neitzel & Welzer, *Soldaten*, p. 207.
30. Hannes Trautloft, diary.
31. Interview with Erich Hartmann.
32. Neitzel & Welzer, p. 45 & 48.
33. Ibid., p. 58.
34. Ibid., p. 63.
35. Interview with Heinz Rökker. See also Thimmig, *Nattens jägare: German nattjaktflygare i World War II*.
36. Interview with Alfred Grislawski.
37. Neitzel & Welzer, p. 156f.
38. Interview with Hans-Ekkehard Bob.
39. The history of Jewish pilots during World War II will be described in detail in Voume 2 of *German Airmen*

40 Neitzel & Welzer, p. 210.
41 Ibid., p. 159.
42 Interview with Hansgeorg Bätcher.
43 Interview with Erhard Jähnert.
44 Interview with Heinz Lange.
45 Hannes Trautloft, diary entries.
46 Interview with Arthur Gärtner.
47 Interview with Adolf Galland.
48 Via Manfred Wägenbaur, Traditionsgeschwader JG 52.
49 Interview with Hans Hahn.
50 Hannes Trautloft, diary entries.
51 Interview with Alfred Grislawski.
52 Jürgen Grislawski, July 2017.
53 Johannes Gentzen reached 18 aerial victories before being killed in an accident during takeoff with an aircraft in May, 1940.
54 Cornwell, *The Battle of France Then and Now*, p. 320ff.
55 Cornwell, p. 432f.
56 Prien et al, *Die Jagdfliegerverbände der Deutschen Luftwaffe 1934 bis 1945, Teil 5*, p. 401ff.
57 Ibid., p. 461.
58 Holmes, *American Eagles*, p. 95.
59 Hannes Trautloft, diary entries.
60 Ibid.
61 Bundesarchiv/Militärarchiv, RL 10/440.
62 Hannes Trautloft, diary entries.
63 Ibid.
64 TsAMO, f. 263. IAP.
65 TsAMO, f. 215. IAD.
66 TsAMO, f. 215. IAD.
67 Kondrat, *Dostalsha namvek nespokoyny*, p. 134.
68 TsAMO, f. 2. GIAP.
69 Interview with Stepan Mikojan.
70 TsAMO, f. 32. GIAP.
71 Hannes Trautloft, diary entries.
72 Interview with Wolfgang Falck.
73 Grunewald & Puschner, *Das katholische Intellektuellenmilieu in Deutschland, seine Presse und seine Netzwerke (1871-1963)*, p. 486.
74 Rigg, Bryan Mark, *Hitlers judiska soldater*, p. 72.
75 Cornwell, p. 95.
76 Royal Air Force Operations Record Book. 11 Group Operations Record Book September 1939—September 1940: 11 Group Intelligence Bulletin No. 12. 24/7/40. National Archives, Kew, AIR 25/197.

77 Royal Air Force Operations Record Book. 11 Group Operations Record Book September 1939—September 1940: 11 Group Intelligence Bulletin No. 12. 24/7/40. National Archives, Kew, AIR 25/197.
78 von Forell, *Mölders und seine Männer*, p. 154.
79 Jones, *Tigrarna*, p. 235.
80 Ibid, p. 259.
81 Werner Mölders, logbook. Via Victor Mölders.
82 Interview with Gerhard Schöpfel.
83 Hannes Trautloft, diary entries, November 3-7, 1940.
84 Hannes Trautloft, diary entries, November 7, 1940.
85 Interview with Adolf Galland.
86 Möldersgutachten. Militärhistorisches Forschungsamt Potsdam. Am 25. August 2004.
87 Interview with Adolf Galland.
88 Interview with Klaus Häberlen.
89 Interview with Johannes Steinhoff.
90 Interview with Eduard Neumann.
91 Tate, *Hans-Joachim Marseille* p. 99.
92 Interview with Eduard Neumann.
93 Shores et al, *A History of the Mediterranean Air War Volume One: North Africa June 1940-January 1942*, pp. 305 & 413.
94 Shores et al, *A History of the Mediterranean Air War Volume One: North Africa June 1940-January 1942*.
95 Gibbes, *You Live but Once*, p. 212.
96 Interview with Eduard Neumann.
97 http://kristenalexanderauthor.blogspot.se/2013/07/clive-robertson-caldwell-australias.html.
98 Morley-Mower, *Messerschmitt Roulette: The Western Desert 1941-1942*, p. 62.
99 Ibid., p. 63.
100 National Archives, HW 5/31. German section: reports of German Army and Air Force High Grade Machine decrypts.
101 Ring, Hans & Christopher Shores, *Luftkampf zwischen Sand und Sonne*, p. 96.
102 Shores et al, *A History of the Mediterranean Air War 1940-1945. Volume One: North Africa June 1940—January 1942*, p. 345ff.
103 Interview with Eduard Neumann.
104 Interview with Eduard Neumann.
105 Shores et al, *A History of the Mediterranean Air War 1940-1945. Volume Two: North Africa February 1942—March 1943*, p. 123f.
106 Ibid.
107 Australian War Memorial. awm.gov.au/index.php/collection/C1422525.
108 Hans Baur i Heaton & Lewis, *The Star of Africa: The Story of Hans Marseille, the Rogue Luftwaffe Ace Who Dominated the WWII Skies*, p. 122.

[109] Heaton & Lewis, p. 124.
[110] Ibid., p. 124.
[111] Ibid., p. 125.
[112] Ibid., p. 127.
[113] Ibid., p. 127.
[114] Ibid., p. 137.
[115] Kurowski, *Hans-Joachim Marseille*, p. 164.
[116] Shores et al, *A History of the Mediterranean Air War 1940-1945. Volume Two: North Africa February 1942—March 1943*, p. 324ff.
[117] Ibid., p. 339ff.
[118] Ibid., p. 344f.
[119] Kurowski, p. 181.
[120] Shores et al, *A History of the Mediterranean Air War 1940-1945. Volume Two: North Africa February 1942—March 1943*, p. 356f.
[121] Kurowski, p. 183.
[122] Interview with Eduard Neumann.
[123] Interview with Eduard Neumann.
[124] Interview with Heinrich Graf von Einsiedel.
[125] Interview with Heinrich Graf von Einsiedel.
[126] Interview with Heinrich Graf von Einsiedel.
[127] Helms, *Von Anfang an dabei*, p. 103.
[128] RAF Hendon. A/C Serial No. W/NR.360043. Section 2B. Individual history Junkers Ju88 R-1 W/NR.360043/PJ876/8475M. Museum Accession number 78/AF/953. Text by Andrew Simpson. Royal Air Force Museum 2013.
[129] *Bild am Sonntag*, 17/1974.
[130] RAF Hendon. A/C Serial No. W/NR.360043. Section 2B. Individual history Junkers Ju88 R-1 W/NR.360043/PJ876/8475M. Museum Accession number 78/AF/953. Text by Andrew Simpson. Royal Air Force Museum 2013.
[131] According to other information, it was a practice flight.
[132] Hill, *The Great Coup*, p. 26.
[133] Ibid., p. 40.
[134] Paus, *Die Hölle von Hamburg*, p. 25; Saunders, *Arrival of Eagles*, p. 128.
[135] Hill, p. 37.
[136] Rigg, *Hitlers judiska soldater*.
[137] National Archives, Kew. Air 27/1087. No 165 Squadron: Operations Record Book.
[138] Saunders, p. 129.
[139] Deutsche Dienststelle (WASt).
[140] Hannes Trautloft, diary entries, May 29, 1942.
[141] Via Hannes Trautloft.
[142] Interview with Hannes Trautloft.
[143] Interview with Pernille Ravnskov.
[144] NARA. Encounter Report, 1st Lieutenant James Wilkinson, 78th FG.

SOME OTHER BOOKS BY THE AUTHOR

www.vaktelforlag.se

BLACK CROSS RED STAR
– AIR WAR OVER THE EASTERN FRONT
VOLUME 4 STALINGRAD TO KUBAN 1942–1943

Christer Bergström
ISBN 978-91-8844-121-8
400 pages
Large format, heavily illustrated

Regarded as the standard work on the air war over the Eastern Front during the Second World War, Christer Bergström's unique Black Cross/Red Star series covers the history of the air war on the Eastern Front in close detail, from the perspectives of both sides. Based on a close study of German and Russian archive material, as well as interviews with a large number of the airmen who participated in this aerial conflict, it has established itself as the main source on the air war on the Eastern Front.

Black Cross/Red Star, Volume 4 covers the air war along the entire Eastern Front during the winter period of 1942–1943 through March 1943, in great detail, with a balance between German and Soviet archive sources etc, and with many first-hand accounts.

ARNHEM 1944 – AN EPIC BATTLE REVISITED
VOLUME 1: TANKS AND PARATROOPERS

Christer Bergström
ISBN 978-91-88441-44-7
400 pages
Hardcover
Available now!

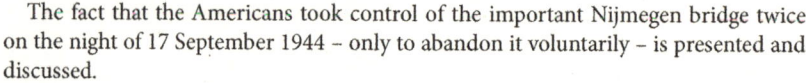

Based upon probably the most profound research into primary sources on "Market Garden" to date, this book will shake the foundations of the prevalent image of this epic battle.

The in-depth research by the author has resulted in many myths and misconceptions being convincingly dispelled, backed up by detailed source notes.

Here, the previously unknown detour that the British paratroopers were forced to take on their way to the Arnhem bridge, costing them absolutely crucial time, is uncovered.

The fact that the Americans took control of the important Nijmegen bridge twice on the night of 17 September 1944 – only to abandon it voluntarily – is presented and discussed.

These and many more similar new facts form a completely new image of Operation "Market Garden". This book is a must-read to anyone interested in this battle, and who does not want to miss out on newly revealed knowledge of this fascinating sequence of World War II.

"Volume One of Arnhem 1944 is a solid hit. Christer Bergström gets high marks for original research, comprehensive orders of battle, and interactive QR codes that immerse the reader into the heat of battle. This puts a whole new face on Market-Garden."
– Andy Nunez, editor of Against the Odds Magazine.

Read more about the book here:

https://vaktelforlag.se/produkt/arnhem-1944-epic-battle-revisited-vol-1-tanks-paratroopers/

A signed copy of the book can be ordered from the author at vaktelforlag@gmail.com
Price US$ 32.00, Postage 1 copy: US$ 10.00

ARNHEM 1944 – AN EPIC BATTLE REVISITED
VOLUME 2: THE LOST VICTORY. SEPTEMBER–OCTOBER 1944

Christer Bergström
ISBN 978-91-8844-145-4
424 pages
Hardcover
Available now!

All previous published accounts of Operation "Market Garden" end the main story with the evacuation of the British airborne troops from Oosterbeek – which obscures the fact that Operation Market Garden at that time was still regarded as, essentially, a great success. It was only because of the following development of events (including the battle at Overloon in October 1944) that the strategic success of Operation Market Garden could not be utilized to end the war before the turn of the year 1944–45. This is a story that has never been told before, and which is described and analyzed in detail in the concluding Volume 2 of Christer Bergström's *Arnhem 1944*.

"Military historian Christer Bergström treats the source material with excellence and puts common images and myths of this battle into question. An extensive source and note list, photos, fact boxes, QR-coded film and sound clips supplement the text of this impressive work in two volumes about operation Market Garden. In summary – brilliant."
– Svensk Bibliotekstjänst (Swedish Library Service), on Volume 2.

Read more about the book here:

https://vaktelforlag.se/produkt/arnhem-1944-epic-battle-revisited-vol-2-lost-victory-september-october-1944/

A signed copy of the book can be pre-ordered from the author at vaktelforlag@gmail.com
Price US$ 32.00, Postage 1 copy: US$ 10.00

STALINGRAD – AN EPIC BATTLE REVISITED

Christer Bergström
ISBN 978-91-88441-56-0
330 pages
Hardcover
Large format
Lavishly illustrated
To be published in 2020

Based on decades of research work in both German and Russian archives, as well as interviews with a large number of key figures and veterans, *Stalingrad – An Epic Battle Revisited* brings our knowledge on this turning point of World War II several big steps forward. It brings forward many hitherto unknown facts and dispels many myths and misconceptions of the battle.

This work on the ground war during the Battle of Stalingrad is the continuation of Christer Bergström's previous book *Operation Barbarossa 1941: Hitler against Stalin* and supplements his Black Cross/Red Star series about the air war on the Eastern Front.

"Christer Bergström is recognized as a prominent expert on the Second World War, not least concerning the Eastern Front. Bergström's book is so extensive, profound, and detailed that I am totally taken aback."
– Kjell E. Genberg in DAST Magazine on Christer Bergström's previous book *Operation Barbarossa 1941: Hitler against Stalin*.

Read more about the book here:

 https://vaktelforlag.se/produkt/stalingrad-epic-battle-revisited/

A signed copy of the book can be pre-ordered from the author at vaktelforlag@gmail.com

DAISY – THE HISTORY OF A C-47/DC-3 IN WORLD WAR II AND THE MEN WHO FLEW IT

Christer Bergström
U.K. edition, ISBN 978-91-88441-51-5
U.S. edition, ISBN 978-91-88441-47-8

This is the thrilling story of a C-47 Skytrain which flew over Normandy and during Operation Market Garden in World War II. The book also goes into detail with the men who flew it, and what they experienced during the war.

Called "Daisy," the aircraft still flies, owned by a Swedish organization, and took part in the parachute drop over Normandy on the 75th Anniversary in 2019.

"Christer Bergström, famous for a long string of military history works, has dived deep into British and American archives to unearth Daisy's military past. The results he has come up with are sensational."
– Lennart Berns, Swedish aviation historian.

THE BATTLE OF BRITAIN
– AN EPIC CONFLICT REVISITED

Christer Bergström
ISBN 978-16-1200-347-4
336 pages
Large format, heavily illustrated

The most thorough, expert examination of the topic ever written. Illustrated throughout with maps and rare photos, plus a colour section closely depicting the aircraft, this work lays out the battle as seldom seen before.

This book contains a large number of dramatic eyewitness accounts, even as it reveals new facts that have altered much of the perception of the battle in the public eye. For example, the twin-engined Messerschmitt Bf 110 was actually a good daytime fighter, and it performed at least as well in this role as the Bf 109 during the battle. The Luftwaffe commander, Hermann Göring, performed far better than his public image has previously indicated.

OPERATION BARBAROSSA 1941 – HITLER AGAINST STALIN

Christer Bergström
ISBN 978-1612004013
336 pages
Large format, heavily illustrated

Based on decades of research work in both German and Russian archives, as well as interviews with a large number of key figures and veterans, Operation Barbarossa brings our knowledge on the war on the Eastern Front several big steps forward. It reveals and dispels many myths and misconceptions including: the myth of mass surrenders by Soviet soldiers; the myth about the vast differences in troop casualties between the two sides; the myth of the Soviet partisans and the myth that it was the Arctic cold that halted the German offensive.

"Christer Bergström, recognized as an internationally leading expert on the Eastern Front during the Second World War, has researched for years in both Russian and German archives and conducted numerous interviews. This is a magnificent book that to a large extent is based on the stories of the soldiers involved in addition to the author's insights into the political and military events. He brings forth many hitherto unknown facts about the war in the East, and in addition to that, the book Operation Barbarossa is extensively illustrated with a large number of images never before published."
- DAST Magazine, 5 July 2016.

BLACK CROSS RED STAR –
AIR WAR OVER THE EASTERN FRONT:
VOLUME 5, THE GREAT AIR BATTLES: KUBAN AND KURSK APRIL-JULY 1943

Christer Bergström
ISBN 978-91-88441-57-7
384 pages
Hardcover
Large format
To be published in 2020

This is the direct continuation of Volume 4 in the Black Cross/Red Star series. This volume covers the air war on the Eastern Front between March/April 1943 and July 1943, with the focus on the great air battles at Kuban and Kursk.

Read more about the book here:

https://vaktelforlag.se/produkt/black-cross-red-star-air-war-eastern-front-volume-5-great-air-battles-kuban-kursk-april-july-1943/

A signed copy of the book can be pre-ordered from the author at vaktelforlag@gmail.com

Index

Ageshin, Sergeant First Class 80
Ahlbrand, Gustaf 15-17, 28
Aniskin, Aleksandr 88
Antisemitism 47, 101, 104, 118, 130, 157, 159, 184-185
Antonov, Sergeant First Class 80
Appel, Adolf 182, 189
Axmann, Artur 156-157
Babkov, Vasilij 88
Bader, Douglas 87
Badger, Ivor James 154
Baklan, Andrej 88
Balandin, Second Lieutenant 85
Baldauf-Mölders/Petzolt, Luise 117, 121-122, 127-128
Baranov, Michail 177, 188
Barkhorn, Gerhard 38-40, 202
Barker, Francis 162-163
Bartels, Heinrich 59
Bätcher, Hansgeorg 11, 21, 34, 48
Batz, Willi 11, 23, 39-40, 202
Baumgartner, Herbert 183
Baur, Hans 155, 157
Beckmann, Helmut 31
Beerenbrock, Franz-Josef 125
Behling, Günther 11, 119-120, 128
Beisswenger, Hans 77, 79, 81, 202
von Below, Nicolaus 156
Berendes, Helmut 11, 42
de la Beresford, Tristram 150-151
Birkner, Joachim 40
Birksted, Kaj 197-198
von Bismarck, Otto 174, 176
Blakeslee, Donald 23
Bob, Hans-Ekkehard 2, 11, 35, 47, 54
Boelcke Oswald 36-37, 216
Boenigk, Ernst 108, 124
Bong, Richard 16
Bormann, Martin 156
Boss, Hugo 42
Botha, Louis Cecil 150-151
Briggs, Tommy 148
Broch, Hugo 11, 142
Brooking, Robert 43-44
Bund Deutscher Offiziere 181
Burney, Henry George 170
Byers, Pat 141-142
Caldwell, Clive 141, 148, 170
Campbell, "Ken" 108, 124
Carson, Kenneth 170
Carson, William 170
Cheshire, Leonard 21
Chislov, Aleksandr 86
Churchill, Winston 110, 157
Clostermann, Pierre 22
Cocker, J. 195
Colquhon, David 186
Concentration camps 43, 47, 120, 157-158, 184
Coussens, Herbert William 154
Cuthbert, Gerald I. 68
Darr, Uffe Børge 199-201
Davydov, Aleksandr 86
Denis, James 137-139, 168
Dickinson, Eric 170
Dold, B. E. 144
Dolgushin, Sergej 88
Donkin, Alexander 170
Drew, Roy 153-154, 170
Duke, Neville 170
Dürkopp, Otto 81
Düttmann, Peter 23, 202
Dyer, Fred 22
Egebjerg, Gunnar Christensen 195
Graf von Einsiedel, Heinrich 11, 173, 179, 188
el-Alamein 159-165, 194
Espenlaub, Albert 146
Ewald, Heinz "Esau" 38, 39
Ewald, Wolfgang 174-176, 178, 188
Fabian, Wolfgang 197
Falck, Wolfgang 11, 92
Fedorenko, K. V. 83
Fischer, Oswald 45, 155
Fjodorov, Fjodor 176
Fonaryov, Konstantin 83
Forward, Ronald 111
François, Armando 161
von Frankenberg und Proschlitz, Egbert 180
Franzisket, Ludwig 139
Frost, Jack 139
Friesicke, Lilli 150
Füllgrabe, Heinrich 53
Funcke, Hans 14, 16, 29
von Galen, Clemens August 118
Galland, Adolf 7, 10-11, 22, 30, 35-37, 46-47, 49, 52, 73, 87, 98, 102, 106, 110, 113, 116-117, 129
Galland, Paul 30
Garam, Michail 88
Gärtner, Arthur 11, 48
Gaymans, H. G. 150
Gebhardt, Martin 113
Gentile, Don 76
Gentzen, Johannes 64
Gestapo 158
Gibbes, Robert 140
Gibson, Guy 21
Glinka, Dmitrij 58
Globocnik, Odilo 157
Gneisenau (German ship) 74, 174
Goebbels, Joseph 35, 155-157
Goebbels, Magda 156
Golding, Cecil 150-151
Göring, Hermann 27, 36, 41, 46, 52, 69, 98, 102, 104-105, 110, 125, 155
Graf, Hermann 24, 27, 53, 92

Grazhdaninov, Pavel 85-86
Grimm, Horst 31
Grislawski, Alfred 11, 20, 24, 33, 42, 46-47, 50, 53, 203
Grislawski, Ilse 23
Grislawski, Jürgen 11, 50
Guderian, Heinz 116
Guernica, bombing of 184
Häberlen, Klaus 11, 131
Hachfeld, Willi 47
Häggberg, Ralf 74
Hahn, Gisela 92
Hahn, Hans "Assi" 61-98, 107, 174, 179, 203
von Hahn, Hans 107-108, 124
Hamburg, bombing of 187
Handrick, Gotthard 63
Harild, Anker Tage 199
Hartigs, Hans 44
Hartmann, Erich 11, 17, 39-40, 45, 56, 92, 202
Hauswirth, Wilhelm 15-17, 28
Helbig, Joachim 11, 21
Heller Joseph 21
Henne, Rudolf 21
Herbert, William 147
Herrmann, Benno 21
Hermann, Hans 182, 189
Hess, Rudolf 183
Hitler, Adolf 26, 40, 42-43, 48, 62-63, 72, 99-102, 105-106, 116-117, 155-157, 181-182, 184
Hitlerjugend 4, 130, 156
Hobbie, Gerhard 121
Hogeback, Hermann 21
Hohenberg, Werner 11, 14-17, 29
Homuth, Gerhard 134-135, 139, 142-150, 152, 169
Hood, Hilary Richard 111
Horn, Peter 198
Hrabak, Dietrich 52, 77, 79, 203
Humphreys, Peter 170
Ihlefeld, Herbert 23-24, 131-133, 203
Jansen, Josef 30
Yershov, L. I. 176
Jobst, Jola 53
Johnson, James E. "Johnnie" 16, 24
Johnson, Robert S. 23
Joplin, Scott 130, 156
Jung, Heinrich 7, 77
Jähnert, Erhard 11, 42, 48
Kahle, Hellmuth 21
Kain, Edgar "Cobber" 107-108, 124
Kantwill, Erich 184-187, 191
Kappler, Herbert 158
Keppler, Gerhard 148
Kesselring, Albert 164
Kholzunov, Aleksej 88
Kholodov, Ivan 88

Kinnen, Karl 29
Kircheis, Erich 111, 114
Kittel, Otto 84, 202
Klawitter, Erich 100, 102, 117, 128
Kleshchov, Ivan 88
Knoke, Heinz 199
Kohl, Stefan 43-44, 51
Kolbe, Georg 121
Kondrat, Yemelyan 81-82
Kosolapov, Filipp 82-83
Kotov, Aleksandr 88
Kovachevich, Arkadij 11, 177
Kozhedub, Ivan 32
Kratz, Hans 182, 189
Kroschinski, Hans-Joachim 57
Kugelbauer, Karl 143-144
Laggai, Willi 29
Lange, Heinz 11, 48
Le Breuilly, Arthur 68
Lees, Kurt 183
Lehwess-Litzmann, Walter 181-182, 189, 199
Leningrad 45, 76, 80, 82
Le Roux, Franciskus 147
Lesckowitz, Martin 28
Leu, Rudolf 170
Leppla, Richard 112
Lichtenstein B/C-radar 184, 187
Lipp, Karl 21
London, bombing of 46
Lotzmann, Manfred 14, 16-17, 28
Lovell, Anthony 111
Lusty, Kenneth 159
Lüty, Gerhard 14, 16, 28
Lützow, Günther 77, 127, 203
Luxton, Clifford 162
Machold, Werner 71
MacRobert, Melville 143
Maiwald, Manfred 15-17, 28
Majorov, Aleksandr 81-82
Malan, Adolphus G. "Sailor" 46, 111
Freiherr von Maltzahn, Günther 101
Markhoff, Hans 14, 16-17, 29
Marsden, J. W. 140
Marseille, Charlotte 129, 146, 158, 163, 170
Marseille, Hans-Joachim 30, 47, 129-166, 168-171, 174, 194, 196, 202
Marseille, Siegfried 21, 29, 129
Martin, Mike 151
Maschel, Johann 48
Mayer, Egon 72-74, 203, 207, 209
McKnight, Willie 66
Meltzer, Karl-Heinz 15-17, 28
Mikoyan, Stepan 11, 87-88, 97
Miller, John Newton 193
Mixa, Otto 28

230

Mölders, Werner 25, 36, 46-47, 63-64, 66, 68, 99-128, 131, 203
Mölders, Victor 11, 100-101, 115, 120, 121
Montgomery, Bernard 160, 194
Morely-Mower, Geoffrey 142
Morrison, Robert 150-151
Moser, Karl Erich 183
Mould, Edward 110-111
Mühlschwein, Heinrich 28
Muir, Vivian 150-151
Müller, Karl 71
Müller, Rudolf 21
Mungo-Park, John Colin 111
Munkert, Karl 29
Munro, Donald 141
Münschow, Günter 28
Mussolini, Benito 102, 134, 158
Müncheberg, Joachim 111, 203
Nationalkomitee Freies Deutschland 178, 182
Neuhoff, Hermann 11, 109, 124
Neumann, Eduard 11, 134, 139, 144, 149-150, 152, 161-162, 164-165, 167, 194
Nordmann, Theodor 54
Normandie 24, 26, 70, 72
Obleser, Friedrich 12, 15-17, 29, 203
Odenhardt, Wilhelm 21
Operation "Reinhard" 157-158
Orton, Newell 108, 124
Osterkamp, Theodor 63-64, 110
Pare, Robin 151
Paris 69, 109
Pas de Calais 70, 72, 87
Pattle, Marmaduke Thomas St John "Pat" 16
Pechaud, Robert 106
Perry, James 108, 124
Petkevich, Sergeant First Class 80
Philipp, Hans 44, 48-49, 58, 202
Pius XI 104
Pohl, Leutnant 139
Pohrt, Werner 28
Pokryshkin, Aleksandr 26, 32, 58
Pomier-Layrargues, René 109
Pompei, Jean 128
Pöttgen, Rainer 151, 164
Quéguiner, Roger 106
Rademacher, Rudolf 82-83, 203
Rall, Günther 12, 14, 16-17, 24, 31, 202
Rathmann, Kurt 28
Ravnskov, Knud Erik 191, 198-199
Ravnskov, Pernille 11, 198
Reich, Karl 28
Reid, Frank 148
Rempel, Edgar 71
Repple, Walter 86
von Richthofen, Manfred 27, 36

Rökker, Heinz 12, 46
Röthke, Siegfried 21
Rott, Friedrich 47
Rommel, Erwin 40, 134, 159-160, 163-164
Roscoe, Arthur 185
Rosen, Eric von 116
Rosenberger, Paul 184-185, 187
Rotterdam, bombing of 65
Rossmann, Edmund 12, 24
Rudel, Hans-Ulrich 21
Sandilands, Noel Milne 145
Saunders, Johnny 143
Saynisch, Walter 31
Scamen, Ben 185
Scharnhorst (German ship) 74
Schellmann, Wolfgang 64-65, 69-70
Schilling, David C. 22
Schlang, Joseph 162
Schmid, Wilhelm 182, 189
Schmitt, Herbert (Heinrich) 183-187, 191
Schroer, Werner 149
Schultze, Norbert 37
Schumacher, Karl 14, 16-17, 28, 40
Seliverstov, Second Lieutenant 80
Serrate radar detection and homing device 186
Sewell, Donald 108, 124
Sharp, Charles 186
Sicherheitsdienst (SD) 159
Skrypnik, Ivan 83
Smith, Launcelot 68, 109
Sobolev, Afanasiy 81-82
Sommer, Poul 194-197
Spicer, Francis 154
Stabnau, Georg 31
Stahlschmidt, Hans-Arnold 148, 149
Stalin, Joseph 61, 88, 93, 97
Stalin, Vasiliy 88-89, 91, 93, 97
Stalin, Svetlana 88
Stalingrad 177, 179.180
Steinhoff, Johannes 12, 24, 30, 133-134, 167, 202
Stotz, Max 77, 79-87, 96, 98, 202
Sullivan, John L. 65-68
Süss, Ernst 46, 53
Swingjugend 157, 174
Taylor, Andy 170
Tenner, Siegfried 29
Tenz, Arthur 121
Terp, Ove 196-197
Thorup, Ejnar 196-197
Tobruk 135, 137, 147, 149
Trautloft, Hannes 12, 49, 52, 76-77, 79-83, 91, 115, 122, 196-197
Tucker, "Tommy" 108, 124
Tulpanov, Sergey 178
Turvey, David 162

Udet, Ernst 5, 120
Ukhov, Valentin 89, 91, 93, 97
Ullmann, Heinz 29
Van Vliet, Cornelius 143-144
Vieck, Carl 115
Vorobyev, Mikhail 85
Wagner, Josef 183
Ward, Derek 152-153
Webster, John Terence 112
Weigelt, Arthur 108, 124
Wells, George 22
Wenzel, Paul 120-121
White, B. D. 145
Wilcke, Wolf-Dietrich 108, 124, 202
Wilkinson, James 200
Windows 187
Winn, John 108, 124
Wittmann, Herbert 21
Wöffen, Antonius 44, 47
Wolf, Albin 79, 81, 202
Wolff, Karl 156-157
Wolfrum, Walter 12, 24, 38, 202
Wykeham-Barnes, Peter 136-137

U.S. units
4th Fighter Group 27
78th Fighter Group 200, 201
340th Bomb Group 21-22
386th Fighter Squadron 43

British and Commonwealth units
1 SAAF Squadron 143, 145-147
3 RAAF Squadron 140, 143
4 SAAF Squadron 151
5 SAAF Squadron 151, 161
14 Squadron 143
21 SAAF Squadron 143
33 Squadron 141-142
41 Squadron 111-113
73 Squadron 107, 124, 136-137, 153-154
74 Squadron 110
92 Squadron 162
112 Squadron 148, 153, 170
145 Squadron 162
185 Squadron 193
213 Squadron 159
237 Squadron 145
242 Squadron 65
250 Squadron 140
257 Squadron 111, 113
450 Squadron 149
451 Squadron 142
607 Squadron 65-66, 68, 109

French units
GC II/5 "Lafayette" 106
GC II/7 109

Italian units
4st Stormo 161-162
9th Gruppo 145
17th Gruppo 145

Soviet units
2nd Guards Fighter Regiment (2nd GIAP) 81
27th IAP 177
32nd Guards Fighter Regiment (32nd GIAP) 88-89
102nd IAD/PVO 176
169nd Fighter Regiment (169th IAP) 85, 88, 91
210th Fighter Wing (210th IAD) 88-89
263rd Fighter Regiment (263rd IAP) 80-81
522nd Fighter Regiment (522nd IAP) 80
629th IAP 176
926th IAP 176

German units
Condor Legion: 4, 64, 99, 102, 104, 180, 184
Ergänzungsjagdgruppe Merseburg 33
Ergänzungsjagdgruppe Ost 198
J 88: 4, 102-103
Jagdfliegerschule 5 131, 196-198
JG 1: 200
JG 2: 20, 64, 68-77, 174
JG 3: 5, 115, 174, 176-179
JG 5: 20
JG 11: 199-201
JG 26: 44-45, 75, 110, 113, 115
JG 27: 30, 44, 47, 134-172, 194-195
JG 51: 110-117, 125-126, 198, 205
JG 52: 15, 17, 23, 28-29, 32-33, 38-40, 49, 133, 205
JG 53: 43, 78, 105, 107-108, 124, 150
JG 54: 2, 4, 48-49, 52, 76-91, 122, 196-197, 205
JG 77: 119
JG 134: 102
JG 231: 64
KG 1: 82-83, 181, 199
KG 2: 48
KG 3: 181
KG 27: 121, 183, 190
KG 51: 4, 48, 180-181
KG 54: 182, 189
KG 77: 180
LG 2: 131
NJG 2: 183
NJG 3: 183
ZG 2: 183